Polygroup Theory
and Related Systems

Polygroup Theory
and Related Systems

Bijan Davvaz

Yazd University, Iran

 **World Scientific**

NEW JERSEY · LONDON · SINGAPORE · BEIJING · SHANGHAI · HONG KONG · TAIPEI · CHENNAI

Published by

World Scientific Publishing Co. Pte. Ltd.

5 Toh Tuck Link, Singapore 596224

USA office: 27 Warren Street, Suite 401-402, Hackensack, NJ 07601

UK office: 57 Shelton Street, Covent Garden, London WC2H 9HE

British Library Cataloguing-in-Publication Data
A catalogue record for this book is available from the British Library.

POLYGROUP THEORY AND RELATED SYSTEMS

Copyright © 2013 by World Scientific Publishing Co. Pte. Ltd.

ISBN 978-981-4425-30-8

Printed in Singapore.

Preface

The theory of groups is the oldest branch of ordinary algebra. The concept of a hypergroup which is a generalization of the concept of a group, first was introduced by Marty. Indeed, hypergroups represent a natural extension of groups. In a group, the composition of two elements is an element, while in a hypergroup, the composition of two elements is a set. Application of hypergroups have mainly appeared in special subclasses. For example, polygroups which are certain subclass of hypergroups are used to study color algebra and combinatorics.

The idea to write this book, and more important the desire to do so, is a direct outgrowth of a course I gave in the department of mathematics at Yazd University. One of my main aims was to present an introduction to this recent progress in the theory of polygroups and related systems. I have tried to keep the preliminaries down to a bare minimum. The book covers most of the mathematical ideas and techniques required in the study of polygroups.

The presented book is composed by five chapters. The first chapter contains a fairly detailed discussion of the basic ideas underlying the theory of groups. The second chapter is about hypergroups, their history and some basic results on some important classes of hypergroups such as complete hypergroups, join spaces and canonical hypergroups. The following chapters are about polygroups. In third chapter, the concept of polygroups and some examples are presented. Then, fundamental relations defined on polygroups, isomorphism theorems, permutation polygroups, polygroup hyperrings, solvable polygroups and nilpotent polygroups are studied. The fourth chapter deal with the concepts of weak algebraic hyperstructures; and the notions of H_v-groups (weak hypergroups) and weak polygroups are studied. In the last chapter, we present some combinatorial aspects of polygroups such as chromatic polygroups.

A large number of people have influenced the writing of this book. I wish to thank Professor Comer, Professor Corsini, Professor De Salvo, Professor Freni, Professor Jantosciak, Professor Leoreanu-Fotea, Professor Spartalis, Professor Vougiouklis, Professor Zahedi and my PhD students who helped me directly or indirectly. The final word belongs to my wife. She deserves an accolade for her patient during the (seemingly interminable) time the book was being written.

Finally may I say that I hope that (most days) you enjoy reading this book as much as I (most days) enjoyed writing it.

Bijan Davvaz
Department of Mathematics
Yazd University, Yazd, Iran

Contents

Chapter 1

A Brief Excursion into Group Theory

1.1 Introduction

The concept of a group is one of the most fundamental in modern mathematics. Group theory can be considered the study of symmetry: the collection of symmetries of some object preserving some of its structure forms a group; in some sense all groups arise this way. Although permutations had been studied earlier, the theory of groups really began with Galois (1811-1832) who demonstrated that polynomials are best understood by examining certain groups of permutations of their roots. Since that time, groups have arisen in almost every branch of mathematics. There are three historical roots of group theory:

(1) The theory of algebraic equations;
(2) Number theory;
(3) Geometry.

Euler, Gauss, Lagrange, Abel and Galois were early researchers in the field of group theory. Galois is honored as the first mathematician linking group theory and field theory, with the theory that is now called Galois theory.

Permutations were first studied by Lagrange (1770, 1771) on the theory of algebraic equations. Lagrange's main object was to find out why cubic equations could be solved algebraically. In studying the cubic, for example, Lagrange assumes the roots of a given cubic equation are x', x'' and x'''. Then, taking $1, w, w^2$ as the cube roots of unity, he examines the expression

$$R = x' + wx'' + w^2 x'''$$

and notes that it takes just two different values under the six permutations of the roots x', x'', x'''. But, he could not fully develop this insight because he viewed permutations only as rearrangements, and not as bijections that

1

can be composed. Composition of permutations does appear in work of Ruffini and Abbati about 1800; in 1815 Cauchy established the calculus of permutations.

Galois found that if $r_1, r_2, \ldots, r_n$ are the n roots of an equation, there is always a group of permutations of the r's such that every function of the roots invariable by the substitutions of the group is rationally known, and conversely, every rationally determinable function of the roots is invariant under the substitutions of the group. Galois also contributed to the theory of modular equations and to that of elliptic functions. His first publication on the group theory was made at the age of eighteen (1829), but his contributions attracted little attention until the publication of his collected papers in 1846.

The number-theoretic strand was started by Euler and taken up by Gauss, who developed modular arithmetic and considered additive and multiplicative groups related to quadratic fields. Indeed, in 1761 Euler studied modular arithmetic. In particular he examined the remainders of powers of a number modulo n. Although Euler's work is, of course, not stated in group theoretic terms he does provide an example of the decomposition of an abelian group into cosets of a subgroup. He also proves a special case of the order of a subgroup being a divisor of the order of the group.

Gauss in 1801 was to take Euler's work much further and gives a considerable amount of work on modular arithmetic which amounts to a fair amount of theory of abelian groups. He examines orders of elements and proves (although not in this notation) that there is a subgroup for every number dividing the order of a cyclic group. Gauss also examined other abelian groups. He looked at binary quadratic forms

$$ax^2 + 2bxy + cy^2 \quad \text{where } a, b, c \text{ are integers.}$$

Gauss examined the behavior of forms under transformations and substitutions. He partitions forms into classes and then defines a composition on the classes. Gauss proves that the order of composition of three forms is immaterial so, in modern language, the associative law holds. In fact Gauss has a finite abelian group and later (in 1869) Schering, who edited Gauss's works, found a basis for this abelian group.

Geometry has been studied for a very long time so it is reasonable to ask what happened to geometry at the beginning of the 19th Century that was to contribute to the rise of the group concept. Geometry had began to lose its *metric* character with projective and non-euclidean geometries being studied. Also the movement to study geometry in n dimensions led to an abstraction in geometry itself. The difference between metric and incidence geometry comes from the work of Monge, his student Carnot and perhaps most importantly the work of Poncelet. Non-euclidean geometry was studied by Lambert, Gauss, Lobachevsky and János Bolyai among others.

Möbius in 1827, although he was completely unaware of the group concept, began to classify geometries using the fact that a particular geometry studies properties invariant under a particular group. Steiner in 1832 studied notions of synthetic geometry which were to eventually become part of the study of transformation groups.

Arthur Cayley and Augustin Louis Cauchy were among the first to appreciate the importance of the theory, and to the latter especially are due a number of important theorems. The subject was popularized by Serret, Camille Jordan and Eugen Netto. Other group theorists of the nineteenth century were Bertrand, Charles Hermite, Frobenius, Leopold Kronecker, and Emile Mathieu.

It was Walther von Dyck who, in 1882, gave the modern definition of a group.

The study of what are now called Lie groups, and their discrete subgroups, as transformation groups, started systematically in 1884 with Sophus Lie; followed by work of Killing, Study, Schur, Maurer, and Cartan. The discontinuous (discrete group) theory was built up by Felix Klein, Lie, Poincaré, and Charles Emile Picard, in connection in particular with modular forms.

Other important mathematicians in this subject area include Emil Artin, Emmy Noether, Sylow, and many others.

1.2 The abstract definition of a group and some examples

Let us consider the set consisting of all the integers $\{0, \pm 1, \pm 2, \ldots\}$. The sum $m + n$ of any two integers m and n is also an integer, and the following two rules of addition hold for any arbitrary integers m, n and p:

(1) $m + n = n + m$,
(2) $(m + n) + p = m + (n + p)$.

Furthermore, for any two given integers m and n, the equation

$$m + x = n$$

has a unique solution $x = n - m$, which is also an integer.

Similar situations often occur in many different fields of mathematics, and they do not necessarily concern only the integers. Consider, for instance, the set of all nonsingular 2×2 matrices (that is, all 2×2 matrices A such that the determinant of A is not zero). Let A and B be any 2×2 matrices,

$$A = \begin{bmatrix} a_1 & a_2 \\ a_3 & a_4 \end{bmatrix} \quad \text{and} \quad B = \begin{bmatrix} b_1 & b_2 \\ b_3 & b_4 \end{bmatrix}.$$

Then, the product of A and B is also a 2×2 matrix. This product is defined as

$$AB = \begin{bmatrix} a_1 b_1 + a_2 b_3 & a_1 b_2 + a_2 b_4 \\ a_3 b_1 + a_4 b_3 & a_3 b_2 + a_4 b_4 \end{bmatrix}.$$

With respect to the binary operation of the multiplication of matrices, we note that, in general, AB is not necessarily equal to BA, but the associative law $(AB)C = A(BC)$ is valid for any three 2×2 matrices A, B and C. If the 2×2 matrix A is nonsingular, then the equation $AX = B$ and $YA = B$ have unique solutions, $X = A^{-1}B$ and $Y = BA^{-1}$, respectively, where X and Y are both 2×2 matrices.

A given set of elements together with an operation satisfying the associative law is said to be a group or to form a group if any linear equation has a unique solution which is in the set. Thus, the totality of nonsingular 2×2 matrices together with multiplication is said to be group, as is the set of all the integers with addition.

We will now state the formal definition of a group.

Definition 1.2.1. Let G be a non-empty set together with a binary operation (usually called multiplication) that assigns to each ordered pair (a, b) of elements of G an element ab in G. We say G is a *group* under this operation if the following two properties are satisfied:

(1) For any three elements a, b and c of G, the associative law holds: $(ab)c = a(bc)$.

(2) For two arbitrary elements a and b, there exists x and y of G which satisfy the equations $ax = b$ and $ya = b$.

The following properties of a group are important.

Theorem 1.2.2.

(2′) There is a unique element e in G such that for all $g \in G$, $ge = eg = g$.

(2″) *For any element $a \in G$, there is a unique element $a' \in G$ such that $aa' = a'a = e$, where e is the element of G defined in (2′).*

(3) *The solutions x and y of the equations $ax = b$ and $ya = b$ are unique and we have $x = a'b$ and $y = ba'$, where a' is the element associated with the element a in (2″).*

Proof. Since the set G is not empty, we take an element a of G. By the property (2), there are solutions $x = e$ and $y = e'$ of the equations $ax = a$ and $ya = a$. Also, if g is an arbitrary element of G, there are elements u and v of G such that $au = g = va$; so we have $ge = (va)e = v(ae) = va = g$. The second equality, $(va)e = v(ae)$, follows from the associative law. Similarly, we obtain $e'g = g$. Since the element g is arbitrary, we may take $g = e$ to obtain $e'e = e$. On the other hand, the element e satisfies $ge = g$ for any $g \in G$, so $e'e = e'$. Therefore, we have $e' = e'e = e$. This proves that an arbitrary solution of the equation $ax = a$ is equal to a solution of $ya = a$. Thus, the uniqueness of the element e is proved, and (2′) holds.

The proof of (2″) is similar. By (2), there are elements a' and a'' of G such that $aa' = e = a''a$. Using (1) and (2′), we have $a'' = a''e = a''(aa') = (a''a)a' = ea' = a'$. Hence, the proof of the uniqueness of a' is similar to that of the uniqueness of e in (2′).

The proof of (3). If $ax = b$, then the left multiplication of a' of (2″) gives us $a'b = a'(ax) = (a'a)x = ex = x$. Thus, a solution of $ax = b$ is $x = a'b$, and it is unique; similarly, the solution of $ya = b$ is uniquely determined to be $y = ba'$. ∎

Corollary 1.2.3. *A non-empty set G with an operation is a group if the conditions (1), (2′) and (2″) are satisfied.*

The element e defined in (2′) is called the *identity* of G, the element a' defined in (2″) is called the *inverse* of a. The inverse of an element a is customary denoted by a^{-1}.

Theorem 1.2.4. *We have $(a^{-1})^{-1} = a$ and $(ab)^{-1} = b^{-1}a^{-1}$, for all $a, b \in G$.*

Proof. The first equality follows from (2″)(the uniqueness of the inverse). The second one is proved by the equality

$$(ab)(b^{-1}a^{-1}) = ((ab)b^{-1})a^{-1} = (a(bb^{-1}))a^{-1} = (ae)a^{-1} = aa^{-1} = e$$

and the uniqueness of the inverse. Notice the change in the order of the factors from ab to $b^{-1}a^{-1}$. ∎

The product of the elements $a_1, a_2, \ldots, a_n$ ($n \geq 3$) of G is defined inductively by $a_1 \ldots a_n = (a_1 \ldots a_{n-1})a_n$. The *general associative law* holds in any group. That is, if $x_1, x_2, \ldots, x_n$ are n arbitrary elements of a group, then the product of $x_1, \ldots, x_n$ is uniquely determined irrespective of the ways the product is taken, provided that the order of factors is unchanged. For example, $(xy)((z(uv))w) = x(((y(zu))v)w)$.

If $a_1 = a_2 = \ldots = a_n$, then we use the power notation (for $a = a_1$), $a_1 a_2 \ldots a_n = a^n$. If $n = -m$ is a negative integer, then we define $a^n = (a^{-1})^m$; also, we define $a^0 = e$. The formulas $a^m a^n = a^{m+n}$ and $(a^m)^n = a^{mn}$ hold for any element a of G and any pair of integers m and n.

We say that two elements a and b of a group G are commutative or commute if $ab = ba$. A group is said to be *abelian* or *commutative*, if any two elements commute.

The number of elements in a group G is called the *order* of G and is denoted by $|G|$. If $|G|$ is finite, then G is said to be a finite group; otherwise G is an infinite group.

Before going on to work out some properties of groups, we pause to examine some examples. Motivated by these examples we shall define various special types of groups which are important.

Example 1.2.5.

(1) The set of integers $\mathbb{Z}$, the set of rational numbers $\mathbb{Q}$ and the set of real numbers $\mathbb{R}$ are all groups under ordinary addition.

(2) The set $\mathbb{Z}_n = \{0, 1, \ldots, n-1\}$ for $n \geq 1$ is a group under addition modulo n. For any i in $\mathbb{Z}_n$, the inverse of i is $n - i$. This group usually referred to as the *group of integers modulo n*.

(3) For a positive integer n, consider the set $C_n = \{a^0, a^1, \ldots, a^{n-1}\}$. On C_n define a binary operation as follows:

$$a^l a^m = \begin{cases} a^{l+m} & \text{if } l + m < n \\ a^{(l+m)-n} & \text{if } l + m \geq n. \end{cases}$$

For every positive integer n, C_n is an abelian group. The group C_n is called the *cyclic group* of order n.

(4) For all integers $n \geq 1$, the set of complex roots of unity

$$\left\{ \cos\frac{2k\pi}{n} + i\sin\frac{2k\pi}{n} \mid k = 0, 1, 2, \ldots, n-1 \right\}$$

(i.e., complex zeros of $x^n - 1$) is a group under multiplication.

(5) In mathematics, a dihedral group is the group of symmetries of a regular polygon, including both rotations and reflections. Dihedral groups are among the simplest examples of finite groups, and they play an important role in group theory, geometry, and chemistry.

(6) The *quaternion group* is a non-abelian group of order 8. It is often denoted by Q or Q_8 and written in multiplicative form, with the following 8 elements

$$Q = \{1, -1, i, -i, j, -j, k, -k\}.$$

Here 1 is the identity element, $(-1)^2 = 1$ and $(-1)a = a(-1) = -a$ for all a in Q. The remaining multiplication rules can be obtained from the following relation:

$$i^2 = j^2 = k^2 = ijk = -1.$$

(7) If n is a positive integer, we can consider the set of all invertible $n \times n$ matrices over the real numbers. This is a group with matrix multiplication as operation. It is called the *general linear group*, $GL(n)$. Geometrically, it contains all combinations of rotations, reflections, dilations and skew transformations of n-dimensional Euclidean space that fix a given point (the origin). If we restrict ourselves to matrices with determinant 1, then we obtain another group, the *special linear group*, $SL(n)$. Geometrically, this consists of all the elements of $GL(n)$ that preserve both orientation and volume of the various geometric solids in Euclidean space. If instead we restrict ourselves to orthogonal matrices, then we obtain the *orthogonal group* $O(n)$. Geometrically, this consists of all combinations of rotations and reflections that fix the origin. These are precisely the transformations which preserve lengths and angles. Finally, if we impose both restrictions, then we obtain the *special orthogonal group* $SO(n)$, which consists of rotations only. These groups are first examples of infinite non-abelian groups.

1.3 Subgroups

The concept of subgroups is one of the most basic ideas in group theory.

Definition 1.3.1. A non-empty subset H of a group G is said to be a *subgroup* of G if the following conditions are satisfied:

(1) $a, b \in H$ implies $ab \in H$;

(2) $a \in H$ implies $a^{-1} \in H$.

Corollary 1.3.2. *If H is a subgroup of G, then H is a group in its own right.*

Corollary 1.3.3. *Let G be a group and H be a non-empty subset of G. Then, H is a subgroup of G if H is closed under division, i.e., if ab^{-1} is in H, whenever a, b are in H.*

Corollary 1.3.4. *Let H be a non-empty finite subset of a group G. Then, H is a subgroup of G if H is closed under the operation of G.*

Example 1.3.5.

(1) Let G be the group of all real numbers under addition, and let H be the set of all integers. Then, H is a subgroup of G.

(2) Let G be the group of all nonzero real numbers under multiplication, and let H be the set of positive rational numbers. Then, H is a subgroup of G.

(3) Let G be the group of all nonzero complex numbers under multiplication, and let $H = \{a + bi \mid a^2 + b^2 = 1\}$. Then, H is a subgroup of G.

(4) Let G be an abelian group. Then, $H = \{x \in G \mid x^2 = e\}$ is a subgroup of G.

(5) Let G be the multiplicative group of all nonsingular 2×2 matrices over complex numbers. Let H be the set of the following eight matrices

$$\pm \begin{bmatrix} 1 & 0 \\ 0 & 1 \end{bmatrix}, \ \pm \begin{bmatrix} i & 0 \\ 0 & -i \end{bmatrix}, \ \pm \begin{bmatrix} 0 & i \\ i & 0 \end{bmatrix}, \ \pm \begin{bmatrix} 1 & 0 \\ 0 & -1 \end{bmatrix}.$$

Then, H is a subgroup of G.

(6) The center $Z(G)$ of a group G is the subset of elements in G that commute with every element of G. In symbols, $Z(G) = \{a \in G \mid ax = xa, \text{ for all } x \in G\}$. The center of a group G is a subgroup of G.

(7) If H is a subgroup of G, then by the *centralizer* $C(H)$ of H we mean the set $\{x \in G \mid xh = hx \text{ for all } h \in H\}$. Then, $C(H)$ is a subgroup of G.

(8) Let $a \in G$, define $N(a) = \{x \in G \mid xa = ax\}$. Then, $N(a)$ is a subgroup of G. $N(a)$ is usually called the *normalizer* or *centralizer* of a in G.

(9) If G is a group and $a \in G$, then the *cyclic subgroup generated by a*, denoted by $< a >$, is the set of all the powers of a.

Among the subgroups of G, the subgroups G and $\{e\}$ are said to be *trivial*. A subgroup H is said to be a *proper subgroup* of G if $H \neq G$. If M is a proper subgroup of G and if $M \subseteq H \subseteq G$ for a subgroup H of G implies that $G = H$ or $H = M$, then M is said to be a *maximal subgroup* of G.

Proposition 1.3.6. *Let H and K be two subgroups of a group G. The intersection $H \cap K$ of H and K is a subgroup of G. In general if $\{H_i\}_{i \in I}$ is a family of subgroups of G, then $\bigcap_{i \in I} H_i$ is a subgroup of G.*

Definition 1.3.7. If X is a subset of a group G, then the smallest subgroup of G containing X, denoted by $< X >$, is called the *subgroup generated by* X. If X consists of a single element a, then $< X >=< a >$, the cyclic subgroup generated by a.

Theorem 1.3.8. *If X is a non-empty subset of a group G, then the subgroup $< X >$ is the set of all finite products of the form $u_1 u_2 \ldots u_n$, where for each i, either $u_i \in X$ or $u_i^{-1} \in X$.*

Proof. Let H be the set of all finite products of the form $u_1 u_2 \ldots u_n$, where u_i or $u_i^{-1} \in X$ and n any positive integer. Consider $x = a_1 a_2 \ldots a_n$ and $y = b_1 b_2 \ldots b_m$ in H. Then, $xy = a_1 a_2 \ldots a_n b_1 b_2 \ldots b_m$ is a product of finite number of elements a_i, b_j such that either the factor or its inverse is in X, consequently $xy \in H$. Further, $x^{-1} = a_n^{-1} \ldots a_2^{-1} a_1^{-1}$. Since a_i or a_i^{-1} is in X, and $a_i = (a_i^{-1})^{-1}$, we see that either a_i^{-1} or $(a_i^{-1})^{-1}$ is in X, and so $x^{-1} \in H$. This proves that H is a subgroup of G. Clearly, $X \subseteq H$. Consider any subgroup K of G containing X. Then, for each $u \in X$, $u \in K$ and hence $u^{-1} \in K$. Thus, if $x = u_1 u_2 \ldots u_n$, where $u_i \in X$ or $u_i^{-1} \in X$, is any element of H, then $x \in K$, since $u_i \in K$ for all i. Hence, $H \subseteq K$. This proves that H is the subgroup of G generated by X. ∎

Definition 1.3.9. Let G be a group and H be a subgroup of G. For $a, b \in G$ we say *a is congruent to b mod H*, written as $a \equiv b \bmod H$ if $ab^{-1} \in H$.

Lemma 1.3.10. *The relation $a \equiv b \bmod H$ is an equivalence relation.*

Definition 1.3.11. If H is a subgroup of G and $a \in G$, then $Ha = \{ha \mid h \in H\}$. Ha is called a *right coset* of H in G. A *left coset* aH is defined similarly. The number of distinct right cosets of H is called the *index* of H in G and denoted by $[G : H]$.

Lemma 1.3.12. *For all $a \in G$, we have $Ha = \{x \in G \mid a \equiv x \bmod H\}$.*

The following corollary contains the basic properties of right cosets and it is useful in many applications.

Corollary 1.3.13. *Let H be a subgroup of G.*

    (1) Every element a of G contained in exactly one coset of H. This coset is Ha.

    (2) Two distinct cosets of H have no common element.

    (3) The group G is partitioned into a disjoint union of cosets of H.

    (4) There is a one to one correspondence between any two right cosets of H in G.

    (5) There is a one to one correspondence between the set of left cosets of H in G and the set of right cosets of H in G.

Theorem 1.3.14. *If G is a finite group and H is a subgroup of G, then $|H|$ is a divisor of $|G|$.*

The above theorem of Lagrange is one of the basic results in finite group theory.

Definition 1.3.15. *If G is a group and $a \in G$, then the order of a is the least positive integer n such that $a^n = e$. If no such integer exists we say that a is of infinite order. We use the notation $o(a)$ for the order of a.*

There are many useful corollaries of Lagrange theorem.

Corollary 1.3.16. *A finite cyclic group of prime order contains no non-trivial subgroup.*

Corollary 1.3.17. *The order of an element of a finite group G divides the order $|G|$.*

Definition 1.3.18. *Let A and B be two subsets of a group G. The set $AB = \{ab \mid a \in A, \ b \in B\}$ consisting of the products of elements $a \in A$ and $b \in B$ is said to be the product of A and B.*

The associative law of multiplication gives us $(AB)C = A(BC)$ for any three subsets A, B and C.

The product of two subgroups is not necessarily a subgroup. We have the following theorem.

Theorem 1.3.19. *Let A and B be subgroups of a group G. Then, the following two conditions are equivalent:*

(1) The product AB is a subgroup of G;

(2) We have $AB = BA$.

Proof. Suppose that AB is a subgroup of G. Then, for any $a \in A$, $b \in B$, we have $a^{-1}b^{-1} \in AB$ and so $ba = (a^{-1}b^{-1})^{-1} \in AB$. Thus, $BA \subseteq AB$. Now, if x is any element of AB, then $x^{-1} = ab \in AB$ and so $x = (x^{-1})^{-1} = (ab)^{-1} = b^{-1}a^{-1} \in BA$, so $AB \subseteq BA$. Thus, $AB = BA$.

On the other hand, suppose that $AB = BA$, i.e., if $a \in A$ and $b \in B$, then $ab = b_1a_1$ for some $a_1 \in A$, $b_1 \in B$. In order to prove that AB is a subgroup we must verify that it is closed and every element in AB has its inverse in AB. Suppose that $x = ab \in AB$ and $y = a'b' \in AB$. Then, $xy = aba'b'$, but since $ba' \in BA = AB$, $ba' = a_2b_2$ with $a_2 \in A$, $b_2 \in B$. Hence, $xy = a(a_2b_2)b' = (aa_2)(b_2b') \in AB$. Clearly, $x^{-1} = b^{-1}a^{-1} \in BA = AB$. Thus, AB is a subgroup of G. ■

1.4 Normal subgroups and quotient groups

There is one kind of subgroup that is especially interesting. If G is a group and H is a subgroup of G, it is not always true that $aH = Ha$ for all $a \in G$. There are certain situations where this does hold, however, and these cases turn out to be of critical importance in the theory of groups. It was Galois, who first recognized that such subgroups were worthy of special attention.

Definition 1.4.1. A subgroup N of a group G is called a *normal subgroup* of G if $aN = Na$ for all $a \in G$.

A group G is said to be *simple* if $G \neq \{e\}$ and G contains no non-trivial normal subgroup.

The only simple abelian groups are $\mathbb{Z}_p$ with p prime.

There are several equivalent formulations of the definition of normality.

Lemma 1.4.2. *Let G be a group and N be a subgroup of G. Then,*

(1) N is normal in G if and only if $a^{-1}na \in N$ for all $a \in G$ and $n \in N$.

(2) N is normal in G if and only if the product of two right cosets of N in G is again a right coset of N in G.

Example 1.4.3.

(1) The center $Z(G)$ of a group is always normal. Indeed, any subgroup of $Z(G)$ is normal in G.

(2) If H has only two left cosets in G, then H is normal in G.

(3) Let G be the set of all real matrices $\begin{bmatrix} a & b \\ 0 & d \end{bmatrix}$ where $ad \neq 0$, under matrix multiplication. Then, $N = \left\{ \begin{bmatrix} 1 & b \\ 0 & 1 \end{bmatrix} \mid b \in \mathbb{R} \right\}$ is a normal subgroup of G.

(4) Let $\mathbb{Q}$ be the set of all rational numbers and $G = \{(a, b) \mid a, b \in \mathbb{Q}, a \neq 0\}$. Define $*$ on G as follows: $(a, b) * (c, d) = (ac, ad + b)$. Then, $(G, *)$ is a non-abelian group. If we consider $N = \{(1, b) \mid b \in \mathbb{Q}\}$, then N is a normal subgroup of G.

(5) For every $n \geq 1$, $SL(n)$ is a normal subgroup of $GL(n)$.

Theorem 1.4.4. *Let N be a normal subgroup of a group G, and let $\overline{G}$ denote the set of all cosets of N. For any two elements X and Y of $\overline{G}$, we define their product XY as the subset of G obtained by taking the product of the two subsets X and Y of G. Then, XY is a coset of N. With respect to this multiplication on $\overline{G}$, the set $\overline{G}$ forms a group.*

Proof. Let X and Y be two elements of $\overline{G}$. Then, there are elements x and y of G such that $X = Nx$ and $Y = Ny$. By assumption, N is normal so that $Nx = xN$ for any $x \in G$. Whence

$$XY = (Nx)(Ny) = N(xN)y = N(Nx)y = Nxy.$$

This proves that XY is a coset of N. If $Z \in \overline{G}$, then by the associative law for the product of subsets, we have $(XY)Z = X(YZ)$. Thus, the multiplication defined on $\overline{G}$ satisfies the associative law. By definition, we obtain $(Nx)(Ny) = Nxy$; so the coset which contains the identity e of G, namely N, is the identity of $\overline{G}$, and the inverse of Nx is the coset Nx^{-1}. Therefore, $\overline{G}$ forms a group. ∎

The group $\overline{G}$ which was defined in the above theorem is called the *quotient group* of G by N and is written $\overline{G} = G/N$. The mapping $x \longrightarrow Nx$ from G into $\overline{G}$ is called the *canonical map*. The order of the quotient group G/N is equal to the index of the normal subgroup N, i.e., $|G/N| = [G : N]$.

Example 1.4.5.

(1) Let $G = \mathbb{Z}_{18}$ and $N = < 6 >$. Then, $G/N = \{0 + N, \ 1 + N, \ 2 + N, \ 3 + N, \ 4 + N, \ 5 + N\}$.

(2) Let G be a group such that $(ab)^p = a^p b^p$ for all $a, b \in G$, where p is a prime number. Let $N = \{x \in G \mid x^{p^m} = e$ for some m depending on $x\}$. Then, N is a normal subgroup of G. If $\overline{G} = G/N$ and if $\overline{x} \in \overline{G}$ is such that $\overline{x}^p = \overline{e}$, then $\overline{x} = \overline{e}$.

We close this section with the following correspondence theorem.

Theorem 1.4.6. *Let N be a normal subgroup of a group G, and let $\overline{G} = G/N$. For any subgroup $\overline{V}$ of $\overline{G}$, there corresponds a subgroup V of G such that*

$$N \subseteq V \quad \text{and} \quad \overline{V} = V/N.$$

The subgroup V consists of those elements of G which are contained in some elements of $\overline{V}$ and it uniquely determined by $\overline{V}$. Thus, between the set $\overline{\mathcal{G}}$ of subgroups of $\overline{G}$ and the set $\mathcal{G}$ of subgroups of G which contain N, there exists a one to one correspondence, $\overline{V} \longleftrightarrow V$.

Proof. It is straightforward. ■

1.5 Group homomorphisms

Let G be a finite group with n elements $a_1, a_2, \ldots, a_n$. A multiplication table for G is the $n \times n$ matrix with i, j entry $a_i * a_j$:

G	a_1	a_2	$\ldots$	a_n
a_1	$a_1 * a_1$	$a_1 * a_2$	$\ldots$	$a_1 * a_n$
a_2	$a_2 * a_1$	$a_2 * a_2$	$\ldots$	$a_2 * a_n$
$\ldots$	$\ldots$	$\ldots$	$\ldots$	$\ldots$
a_n	$a_n * a_1$	$a_n * a_2$	$\ldots$	$a_n * a_n$

Informally, we say that we "know" a finite group G if we can write a multiplication table for it.

In this section, we consider one of the most fundamental notions of group theory-"homomorphism". The homomorphism term comes from the Greek words "homo", which means like and "morphe", which means form. In our presentation about groups we see that one way to discover information about a group is to examine its interaction with other groups using homomorphisms. A group homomorphism preserves the group operation.

Let us consider two almost trivial examples of groups. Let G be the group whose elements are the numbers 1 and -1 with operation multiplication and let $\hat{G}$ be the additive group $\mathbb{Z}_2$. Compare multiplication tables of these two groups:

$$
\begin{array}{c|cc}
G & 1 & -1 \\
\hline
1 & 1 & -1 \\
-1 & -1 & 1
\end{array}
\qquad
\begin{array}{c|cc}
\hat{G} & 0 & 1 \\
\hline
0 & 0 & 1 \\
1 & 1 & 0
\end{array}
$$

It is quite clear that G and $\hat{G}$ are distinct groups; on the other hand, it is equally clear that there is no significant difference between them. Let us make this idea precise.

Definition 1.5.1. A function f defined on a group G to a group $\hat{G}$ (not necessarily distinct from G) is said to be a (*group*) *homomorphism* from G into $\hat{G}$ if $f(xy) = f(x)f(y)$ for all $x, y \in G$. If f is a surjective homomorphism, i.e., $f(G) = \hat{G}$, then $\hat{G}$ is said to be *homomorphic* to G. If f is a surjective and one to one homomorphism, then f is called an *isomorphism* from G onto $\hat{G}$. If there is an isomorphism from G onto $\hat{G}$, we say that G and $\hat{G}$ are *isomorphic* and write $G \cong \hat{G}$.

Let f be a homomorphism from G into $\hat{G}$. The subset

$$
H = \{x \in G \mid f(x) \text{ is the identity of } \hat{G}\}
$$

is called the *kernel* of f and is denoted by $Ker f$.

Examples 1.5.2.

(1) Every canonical mapping is a homomorphism.
(2) Let G be the group of all positive real numbers under the multiplication of the real numbers and let $\hat{G}$ be the group of all real numbers under addition. Let $f : G \longrightarrow \hat{G}$ be defined by $f(x) = log_{10}x$ for all $x \in G$. Since $log_{10}(xy) = log_{10}x + log_{10}y$, we have $f(xy) = f(x) + f(y)$, so f is a homomorphism. Also, it happens to be onto and one to one.
(3) Let $GL(n)$ be the multiplicative group of all nonsingular $n \times n$ matrices over the real numbers. Let $\mathbb{R}^*$ be the multiplicative group of all nonzero real numbers. We define $f : G \longrightarrow \mathbb{R}^*$ by $f(A) = det A$ for all $A \in GL(n)$. Since for any two $n \times n$ matrices A, B, $det(AB) = det A \cdot det B$, we obtain $f(AB) = f(A)f(B)$. Hence, f is a homomorphism of $GL(n)$ into $\mathbb{R}^*$. Also, f is onto.
(4) Let D_{2n} be the dihedral group defined as the set of all formal symbols $a^i b^j$, $i = 0, 1$, $j = 0, 1, \ldots, n-1$, where $a^2 = e$, $b^n = e$ and $ab = b^{-1}a$. Then, the subgroup $N = \{e, b, b^2, \ldots, b^{n-1}\}$ is normal in G and $D_{2n}/N \cong H$, where $H = \{1, -1\}$ is the group under the multiplication of the real numbers.

Proposition 1.5.3. *Let f be a homomorphism from a group G into a group $\hat{G}$. The following propositions hold:*

(1) $f(e) = e'$, the identity element of $\hat{G}$.

(2) $f(x^{-1}) = f(x)^{-1}$ for all $x \in G$.

(3) The kernel of f is a normal subgroup of G.

(4) Let H be a subgroup of G. The image $f(H) = \{f(x) \mid x \in H\}$ is a subgroup of $\hat{G}$. For a subgroup $\hat{H}$ of $\hat{G}$, the inverse image

$$f^{-1}(\hat{H}) = \{x \in G \mid f(x) \in \hat{H}\}$$

is a subgroup of G.

(5) For two elements x and y of G, $f(x) = f(y)$ if and only if x and y lie in the same coset of the kernel f. In particular, if f is surjective, then f is an isomorphism if and only if the kernel of f is $\{e\}$.

(6) If H is a normal subgroup of G, then $f(H)$ is a normal subgroup of $f(G)$.

We are in a position to establish an important connection between homomorphisms and quotient groups. Many authors prefer to call the next theorem the Fundamental theorem of group homomorphism.

Theorem 1.5.4. *Let f be a homomorphism from a group G onto a group $\hat{G}$. Then, there exists an isomorphism g from $G/\mathrm{ker}f$ onto $\hat{G}$ such that $f = g\varphi$, where φ is the canonical homomorphism from G onto $G/\mathrm{ker}f$. In this case, we say the following diagram is commutative.*

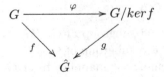

Proof. Suppose that $K = \mathrm{ker}f$. Define a function g from G/K to $\hat{G}$ by

$$g(Kx) = f(x).$$

The function g is well defined and does not depends on the choice of a representative from Kx. Since f is a homomorphism, we have

$$g(KxKy) = g(Kxy) = f(xy) = f(x)f(y) = g(Kx)g(Ky).$$

Hence, g is a homomorphism from G/K onto $\hat{G}$. If the coset Kx lies in the kernel of g, then $e = g(Kx) = f(x)$, so that $x \in K$. Thus, g is an isomorphism. It is clear that $f = g\varphi$ holds; so the diagram is commutative. ∎

Corollary 1.5.5. *Let N be a normal subgroup of a group G, and let φ be the canonical homomorphism from G onto G/N. Let f be a homomorphism from G into a group $\hat{G}$. Then, there exists a homomorphism g from G/N into $\hat{G}$ such that $f = g\varphi$ if and only if $N \subseteq \ker f$. In this case, we have $f(G) \cong (G/N)/(\ker f/N)$.*

Proof. If there is a homomorphism g satisfying $f = g\varphi$, we have $f(N) = e$; so $N \subseteq \ker f$. Conversely, suppose that $N \subseteq \ker f$. Set $K = \ker f$ and $\overline{G} = G/N$. We define a function g from $\overline{G}$ into $\hat{G}$ by

$$g(Nx) = f(x).$$

The function g is uniquely determined independent of the choice of a representative x from Nx, and g is a homomorphism from $\overline{G}$ into $\hat{G}$. Clearly, $f = g\varphi$ by definition, and we have $\ker g = \overline{K}$. From Theorem 1.5.4, we get

$$G/K = f(G) \cong g(\overline{G}) \cong \overline{G}/\overline{K} = (G/N)/(K/N). \blacksquare$$

Corollary 1.5.6. *Let f be a homomorphism from a group G onto a group $\hat{G}$. Let $\hat{N}$ be a normal subgroup of $\hat{G}$ and set $N = f^{-1}(\hat{N})$. Then, N is a normal subgroup of G and $G/N \cong \hat{G}/\hat{N}$.*

Theorem 1.5.7. *Let H be a normal subgroup of a group G, and let K be any subgroup of G. Then, the following hold:*

(1) HK is a subgroup of G;
(2) $H \cap K$ is a normal subgroup of K;
(3) $HK/H \cong K/H \cap K$.

Proof. Since H is a normal subgroup of G, $HK = KH$ and so according to Theorem 1.3.19, HK is a subgroup of G.

Now, consider the canonical mapping from G onto the factor group G/H. Let f be the restriction of the canonical mapping on K. Then, f is a homomorphism from K into G/H. By definition, f is a homomorphism from K onto HK/H. Theorem 1.5.4 proves that

$$HK/H \cong K/\ker f.$$

But, the kernel of the canonical homomorphism is K, whence we get $\ker f = H \cap K$. This completes the proof. $\blacksquare$

Example 1.5.8.

(1) Any finite cyclic group of order n is isomorphic to $\mathbb{Z}_n$. Any infinite cyclic group is isomorphic to $\mathbb{Z}$.

(2) Let G be the group of all real valued functions on the unit interval $[0, 1]$, where for any $f, g \in G$, we define addition $f + g$ by

$$(f + g)(x) = f(x) + g(x) \quad \text{for all } x \in [0, 1].$$

If $N = \{f \in G \mid f(\frac{1}{4}) = 0\}$, then $G/N \cong \mathbb{R}$ under $+$.

(3) The quotient group $\mathbb{R}/\mathbb{Z}$ is isomorphic to the group S^1 of complex numbers of absolute value 1 (with multiplication). An isomorphism is given by $f(x + \mathbb{Z}) = e^{2\pi x i}$ for all $x \in \mathbb{R}$.

(4) The quotient groups and subgroups of a cyclic groups are cyclic.

Definition 1.5.9. By an *automorphism* of a group G we shall mean an isomorphism of G onto itself. If g is an element of G, then the function $i_g : x \mapsto g^{-1}xg$ is an automorphism of G, which is called the *inner automorphism* by g.

Let $Aut(G)$ denote the set of all automorphisms of G. For the product of elements of $Aut(G)$ we can use the composition of mappings.

Lemma 1.5.10. *If G is a group, then $Aut(G)$ is also a group.*

$Aut(G)$ is called the *group of automorphism* of G. The set of all inner automorphisms is a subgroup of $Aut(G)$, which is written $Inn(G)$, and is called the *group of inner automorphism* of G. If G is abelian, then $Inn(G) = \{e\}$.

Theorem 1.5.11. *The group of inner automorphisms of G is isomorphic to the quotient group $G/Z(G)$, where $Z(G)$ is the center of G. Furthermore, $Inn(G)$ is a normal subgroup of $Aut(G)$.*

Proof. The function $g \mapsto i_g$ is a homomorphism from G onto $Inn(G)$. Thus, $G/K \cong Inn(G)$, where K is the kernel of the above homomorphism. An element g of G lies in the kernel K if and only if $i_g = e$, i.e., $g^{-1}xg = x$, for all $x \in G$. Therefore, we have $K = Z(G)$.

The last assertion follows immediately from the formula

$$\sigma^{-1}i_g\sigma = i_{\sigma(g)},$$

which holds for any $g \in G$ and $\sigma \in Aut(G)$. ∎

Example 1.5.12.

(1) $Aut(\mathbb{Z}) \cong \mathbb{Z}_2$;
(2) $Aut(G) = 1$ if and only if $|G| \leq 2$.

1.6   Permutation groups

In this section, we study certain group of functions called permutation group, from a set X to itself. Permutation groups are an important class of groups.

Definition 1.6.1. Let X be a set. A one to one function from X onto X is called a *permutation* on the set X. We denote the set of all permutations of X by $\mathbb{S}_X$.

A function p defined on the set X is a permutation if and only if the following three conditions are satisfied:

(1) $p(x) \in X$ for all $x \in X$;
(2) Given $a \in X$, there is an element $x \in X$ such that $p(x) = a$;
(3) $p(u) = p(v)$ implies $u = v$.

In the important special case when $X = \{x_1, x_2, \ldots, x_n\}$, we write $\mathbb{S}_n$ instead of $\mathbb{S}_X$. Note that $|\mathbb{S}_n| = n!$. If $f \in \mathbb{S}_n$, then f is a one to one mapping of X onto itself, and we could write f out by showing what it does to every element, e.g., $f : x_1 \longrightarrow x_2,\ x_2 \longrightarrow x_4,\ x_4 \longrightarrow x_3,\ x_3 \longrightarrow x_1$. But this is very cumbersome. One short cut might be to write f out as

$$\begin{pmatrix} x_1 & x_2 & x_3 & \ldots & x_n \\ x_{i_1} & x_{i_2} & x_{i_3} & \ldots & x_{i_n} \end{pmatrix},$$

where x_{i_k} is the image of x_k under f. Return to our example just above, f might be represented by

$$\begin{pmatrix} x_1 & x_2 & x_3 & x_4 \\ x_2 & x_4 & x_1 & x_3 \end{pmatrix}.$$

While this notation is a little handier there still is waste in it, for there seems to be no purpose served by the symbol x. We could equally well present the permutation as

$$\begin{pmatrix} 1 & 2 & 3 & \ldots & n \\ i_1 & i_2 & i_3 & \ldots & i_n \end{pmatrix}.$$

If p and q are permutations on a set X, the composition $p \circ q$ is also a permutation on X which is defined to be the *product* of p and q. We write the product of p and q as the juxtaposition pq. For example, $p = \begin{pmatrix} 1 & 2 & 3 \\ 3 & 2 & 1 \end{pmatrix}$ and $q = \begin{pmatrix} 1 & 2 & 3 \\ 2 & 3 & 1 \end{pmatrix}$ are permutations of $\{1, 2, 3\}$. The product pq is $\begin{pmatrix} 1 & 2 & 3 \\ 2 & 1 & 3 \end{pmatrix}$.[1]

---

[1]Notice that some authors compute this product in the reverse order. This authors will write functions on the right: instead of $f(x)$, they write $(x)f$.

The set $\mathbb{S}_X$ of permutations on X forms a group under the operation defined above. We call $\mathbb{S}_X$ the *symmetric group* on X.

Definition 1.6.2. Let X be a set and $\mathbb{S}_X$ be the symmetric group on X. Any subgroup of $\mathbb{S}_X$ is called a *permutation group* on X.

If $x \in X$ and $f \in \mathbb{S}_X$, then f *fixes* i if $f(i) = i$ and f *moves* i if $f(i) \neq i$. We say $\sigma_1, \sigma_2, \ldots, \sigma_k$ are *disjoint* if no element of X is moved by more than one of $\sigma_1, \sigma_2, \ldots, \sigma_k$. It is easy to see that $\tau\sigma = \sigma\tau$ whenever σ and τ are disjoint.

Definition 1.6.3. Let f be a permutation on the set $\{1, 2, \ldots, n\}$ and let $i_1, i_2, \ldots, i_r$ $(1 \leq r \leq n)$ be distinct elements of X. If f fixes the remaining $n - r$ integers and if

$$f(i_1) = i_2, \ f(i_2) = i_3, \ldots, f(i_{r-1}) = i_r, \ f(i_r) = i_1,$$

then f is called a *cycle permutation* or *cycle* and is denoted briefly $(i_1 \ i_2 \ \ldots i_r)$. r is called the *length* of the cycle.

We multiply cycles by multiplying the permutations they represent.

Lemma 1.6.4. *Every permutation can be uniquely expressed as a disjoint cycles.*

A cycle of length 2 is called a *transposition*.

Lemma 1.6.5. *Every permutation is a product of transpositions.*

Let the permutation f on the set $X = \{1, 2, \ldots, n\}$ be expressible in some way as an even (respectively odd) number of transpositions. Then, every way of expressing f as a product of transpositions requires an even (respectively odd) number of transpositions.

Definition 1.6.6. A permutation $f \in \mathbb{S}_n$ is said to be an even permutation if it can be represented as a product of an even number of transpositions. We call a permutation *odd* if it is not an even permutation.

The rule for combining even and odd permutations is like that of combining even and odd numbers under addition.

Let A_n be the subset of $\mathbb{S}_n$ consisting of all even permutations. Since the product of two even permutation is even, A_n must be a subgroup of $\mathbb{S}_n$. A_n is called the *alternative group* of degree n.

Lemma 1.6.7. A_n *is a normal subgroup of* $\mathbb{S}_n$ *of order* $\frac{n!}{2}$.

The alternative groups are among the most important examples of groups.

Lemma 1.6.8. *For $n > 3$, each element of A_n is a product of cycles of length 3.*

Proof. Suppose that $\sigma \in A_n$. Then, σ is an even permutation. Let $(x\ y)$ and $(a\ b)$ be two transpositions. If $\{a, b\} \cap \{x, y\} = \emptyset$, then

$$(a\ b)(x\ y) = (a\ b)(a\ x)(x\ a)(x\ y)$$
$$= [(a\ b)(a\ x)][(x\ a)(x\ y)] = (x\ b\ a)(y\ a\ x).$$

On the other hand, if $\{a, b\} \cap \{x, y\} \neq \emptyset$, then we let $b = x$. So, $(a\ b)(b\ y) = (y\ a\ b)$. ∎.

Lemma 1.6.9. *Let N be a normal subgroup of A_n $(n \geq 5)$. If N contains a cycle of length 3, then $N = A_n$.*

Proof. By Lemma 1.6.8, it is enough to show that each cycle of length 3 is belong to N. Suppose that $(a\ b\ c)$ is an arbitrary cycle of length 3 and $(x\ y\ z) \in N$. Consider $\sigma \in \mathbb{S}_n$ such that $\sigma(a) = x$, $\sigma(b) = y$ and $\sigma(c) = z$. Then, $\sigma^{-1}(x\ y\ z)\sigma = (a\ b\ c)$. If $\sigma \in A_n$, then $(a\ b\ c) \in N$. If $\sigma \notin A_n$, then we can choose u, v distinct from x, y, z. Since $\sigma \notin A_n$, σ is an odd permutation. So, $\sigma(u\ v) \in A_n$. Thus, $((u\ v)\sigma)^{-1}(x\ y\ z)((u\ v)\sigma) \in N$. But

$$((u\ v)\sigma)^{-1}(x\ y\ z)((u\ v)\sigma) = \sigma^{-1}(u\ v)(x\ y\ z)(u\ v)\sigma = \sigma^{-1}(x\ y\ z)\sigma.$$

Hence, $\sigma^{-1}(x\ y\ z)\sigma \in N$ which implies that $(a\ b\ c) \in N$. ∎

Lemma 1.6.10. *Let N be a normal subgroup of A_n $(n \geq 5)$. If N contains a product of two disjoint transpositions, then $N = A_n$.*

Proof. Suppose that $(x\ y)(a\ b) \in N$ such that $\{x, y\} \cap \{a, b\} = \emptyset$. Since $n \geq 5$, we choose z distinct from x, y, a, b. Now, we set $\sigma = (a\ b\ z)$. Clearly, $\sigma \in A_n$. So, $\sigma^{-1}(x\ y)(a\ b)\sigma = (z\ b\ a)(x\ y)(a\ b)(a\ b\ z) \in N$. This implies that $(a\ z)(x\ y) \in N$. Since N is a subgroup, $(x\ y)(a\ b)(a\ z)(x\ y) \in N$. Thus, $(a\ z\ b) \in N$. Hence, N contains a cycle of length 3, and so by Lemma 1.6.9, $N = A_n$. ∎

Theorem 1.6.11. *A_n is simple for all $n \geq 5$.*

Proof. Suppose that $n \geq 5$ and N is a non-trivial normal subgroup of A_n. We show that $N = A_n$ by considering the possible cases.

    (1) N contains a cycle of length 3; hence $N = A_n$ by Lemma 1.6.9.

    (2) N contains an element σ, the product of disjoint cycles, at least

one of which has length $r \geq 4$. Then, $\sigma = (a_1 \ a_2 \ \ldots a_r)\tau$ (disjoint). Let $\delta = (a_1 \ a_2 \ a_3) \in A_n$. Then,

$$\sigma^{-1}(\delta\sigma\delta^{-1}) = \tau^{-1}(a_r \ a_{r-1} \ldots a_1)(a_1 \ a_2 \ a_3)(a_1 \ a_2 \ldots a_r)\tau(a_3 \ a_2 \ a_1)$$
$$= (a_1 \ a_3 \ a_r) \in N.$$

Hence, $N = A_n$ by Lemma 1.6.9.

(3) N contains an element σ, the product of disjoint cycles, at least two of which have length 3, so that $\sigma = (a_1 \ a_2 \ a_3)(a_4 \ a_5 \ a_6)\tau$ (disjoint). Let $\delta = (a_1 \ a_2 \ a_4) \in A_n$. Then as above, N contains $\sigma^{-1}(\delta\sigma\delta^{-1}) = (a_1 \ a_4 \ a_2 \ a_6 \ a_3)$. Hence, $N = A_n$ by case (2).

(4) N contains an element σ that is the product of one cycle of length 3 and some transpositions, say $\sigma = (a_1 \ a_2 \ a_3)\tau$ (disjoint), with τ a product of disjoint transpositions. Then, $\sigma^2 = (a_1 \ a_3 \ a_2) \in N$, so $N = A_n$ by Lemma 1.6.9.

(5) Each element of N is the product of an even number of disjoint transpositions. Let $\sigma \in N$, with $\sigma = (a_1 \ a_2)(a_3 \ a_4)\tau$ (disjoint). Let $\delta = (a_1 \ a_2 \ a_3) \in A_n$. Then as above, N contains $\sigma^{-1}(\delta\sigma\delta^{-1})$. Now, $\sigma^{-1}(\delta\sigma\delta^{-1}) = (a_1 \ a_3)(a_2 \ a_4)$. Since $n \geq 5$, there is an element $b \in \{1, \ldots, n\}$ distinct from a_1, a_2, a_3, a_4. Since $\eta = (a_1 \ a_3 \ b) \in A_n$ and $\zeta = (a_1 \ a_3)(a_2 \ a_4) \in N$, we have $\zeta(\eta\zeta\eta^{-1}) \in N$. But $\zeta(\eta\zeta\eta^{-1}) = (a_1 \ a_3 \ b) \in N$. Hence, $N = A_n$ by Lemma 1.6.9.

Since the cases listed cover all the possibilities, A_n has no proper normal subgroups and hence is simple. ∎

The English mathematician Cayley first noted that every group could be realized as a subgroup of $\mathbb{S}_X$ for some X.

Theorem 1.6.12. (Cayley's theorem). *Let G be a given group. Then, there exists a set X such that G is isomorphic to a permutation group on X.*

Proof. We choose the set X consisting of all the elements of G. For each element $g \in G$, let π_g be a function on X defined by the formula

$$\pi_g(x) = xg \quad (\text{for } x \in X).$$

It follows that π_g is a permutation on X. Furthermore, the associative law proves

$$\pi_{gh} = \pi_g\pi_h \quad (\text{for } g, h \in G).$$

Thus, the function π is a homomorphism from G into the symmetric group $\mathbb{S}_X$. Clearly, π_g is the identity function on X if and only if $g = e$. This means that π is one to one. Hence, G is isomorphic to the image $\pi(G)$ which is a permutation group on X. ∎

1.7   Direct product

There are many methods of constructing new groups from one or more given groups. One of the most important methods is the method of formation of direct product of groups. In this section, we discuss the concepts of direct and semidirect products of groups.

Definition 1.7.1. If H and K are groups, then their *direct product*, denoted by $H \times K$, is the group with elements all ordered pairs (h, k), where $h \in H$ and $k \in K$, and with the operation

$$(h, k)(h', k') = (hh', kk').$$

It is easy to check that $H \times K$ is a group; the identity is (e, e'); the inverse $(h, k)^{-1}$ is (h^{-1}, k^{-1}). Notice that neither H nor K is a subgroup of $H \times K$, but $H \times K$ does contain isomorphic replicas of each, namely, $H \times \{e'\} = \{(h, e') \mid h \in H\}$ and $\{e\} \times K = \{(e, k) \mid k \in K\}$.

Example 1.7.2.

(1) Let $V = \{e, a, b, c\}$ under a binary operation defined by the following table:

*	e	a	b	c
e	e	a	b	c
a	a	e	c	b
b	b	c	e	a
c	c	b	a	e

Then, $(G, *)$ is a group. This is known as *Klein's four group*. We have $V \cong \mathbb{Z}_2 \times \mathbb{Z}_2$.

(2) If $(m, n) = 1$, then $\mathbb{Z}_m \times \mathbb{Z}_n \cong \mathbb{Z}_{mn}$.

Theorem 1.7.3. *Let G be a group with normal subgroups H and K. If $HK = G$ and $H \cap K = \{e\}$, then $G \cong H \times K$.*

Proof. If $a \in G$, then $a = hk$ for some $h \in H$ and $k \in K$. We claim that h and k are uniquely determined by a. If $a = h_1 k_1$ for $h_1 \in H$ and $k_1 \in K$, then $hk = h_1 k_1$ and $h^{-1} h_1 = k k_1^{-1} \in H \cap K = \{e\}$. Hence, $h = h_1$ and $k = k_1$. We define $f : G \longrightarrow H \times K$ by $f(a) = (h, k)$, where $a = hk$. If $a = hk$ and $a' = h'k'$, then $aa' = hkh'k'$ which is not in the proper form for evaluating f. We consider $h'kh'^{-1}k^{-1}$. Then, $(h'kh'^{-1})k^{-1} \in K$, since K is normal. Similarly, $h'(kh'^{-1}k^{-1}) \in H$, since H is normal. Therefore, $h'kh'^{-1}k^{-1} \in H \cap K = \{e\}$ and h' and k commute.

Therefore, $f(aa') = f(hh'kk') = (hh', kk') = (h, k)(h', k') = f(a)f(a')$. Now, it is easy to see that f is an isomorphism. ∎

Theorem 1.7.4. *If A is a normal subgroup of H and B is normal subgroup of K, then $A \times B$ is a normal subgroup of $H \times K$ and*

$$(H \times K)/(A \times B) \cong (H/A) \times (K/B).$$

Proof. The homomorphism $f : H \times K \longrightarrow (H/A) \times (K/B)$, defined by $f(h, k) = (Ah, Bk)$, is surjective and $\ker f = A \times B$. The fundamental homomorphism theorem now gives the result. ∎

It follows, in particular, that if N is normal subgroup of H, then $N \times \{e\}$ is a normal subgroup of $H \times K$.

Corollary 1.7.5. *If $G = H \times K$, then $G/(H \times \{e\}) \cong K$.*

Definition 1.7.6. Let G and H be two groups. If a homomorphism φ from G into $Aut H$ is given, we say that G *acts* on H via φ and G is an *operator group* on H; the homomorphism φ is called an *action* of G.

The action φ of G is not necessarily an isomorphism. In particular, φ can be the trivial action, i.e., $\varphi(g) = i$ for all $g \in G$. Via the trivial action, any group can act on H.

For any $g \in G$, $\varphi(g)$ is an automorphism of H. We denote the image $\varphi(g)(h)$ of an element $h \in H$ simply by h^g. The action on H is given if and only if the function

$$\varphi(g) : h \mapsto h^g$$

are defined for each $g \in G$ and satisfy the following formulas for all $u, v \in H$ and $x, y \in G$:

$$(uv)^x = u^x v^x, \quad u^{xy} = (u^x)^y, \quad u^e = u,$$

where e is the identity of G.

Theorem 1.7.7. *Let φ be an action of a group G on another group H. Let L be the cartesian product of G and H, i.e., the totally of pairs (g, h) of elements $g \in G$ and $h \in H$. We define the product of two elements of L by the formula*

$$(g, h)(g', h') = (gg', \ \varphi(g')(h)h').$$

Then, L forms a group with respect to this operation.

Let e_G denotes the identity of G; similarly, let e_H denotes the identity of H. We define for $g \in G$ and $h \in H$,

$$\gamma(g) = (g, e_H), \quad \eta(h) = (e_G, h).$$

We set

$$\overline{G} = \{\gamma(g) \mid g \in G\}, \quad \overline{H} = \{\eta(h) \mid h \in H\}.$$

Then, γ is an isomorphism from G onto $\overline{G}$, η is an isomorphism from H onto $\overline{H}$, and we have $\overline{H}$ is a normal subgroup of $L = \overline{G}\,\overline{H}$ and $\overline{G} \cap \overline{H} = \{e\}$, where $e = (e_G, e_H)$ is the identity of L. Furthermore, the formula

$$\gamma(g)^{-1}\eta(h)\gamma(g) = \eta(h^g)$$

holds for any $g \in G$ and $h \in H$.

Proof. First, we verify the associative law. Let $g, u, x \in G$ and $h, v, y \in H$. Then,

$$
\begin{aligned}
[(g,h)(u,v)](x,y) &= (gu, h^u v)(x,y) \\
&= ((gu)x, (h^u v)^x y) = (g(ux), ((h^u)^x v^x)y) \\
&= (g(ux), h^{ux}(v^x y)) = (g,h)(ux, v^x y) \\
&= (g,h)[(u,v)(x,y)].
\end{aligned}
$$

By definition, $e = (e_G, e_H)$ is the identity; the inverse of (g,h) is given by $(g^{-1}, \varphi(g^{-1})(h)^{-1})$. So, L forms a group with respect to the operation defined above. The definition of the operation also shows that

$$\gamma(gg') = \gamma(g)\gamma(g'), \quad \eta(hh') = \eta(h)\eta(h');$$

so both γ and η are homomorphisms and $\overline{G} = \gamma(G)$ and $\overline{H} = \eta(H)$ are subgroups. Furthermore, by definition we have

$$
\begin{aligned}
(g,h) &= \gamma(g)\eta(h), \\
\gamma(g)^{-1}\eta(h)\gamma(g) &= \eta(h^g).
\end{aligned}
$$

Hence, $\overline{H}$ is a normal subgroup of $L = \overline{G}\,\overline{H}$. Clearly, we have $\overline{G} \cap \overline{H} = \{e\}$. ∎

The group constructed in the above theorem is called the *semidirect product* of G and H with respect to the action φ.

1.8   Solvable and nilpotent groups

The goal of this section is to introduce the concept of solvable and nilpotent groups.

Definition 1.8.1. A sequence of subgroups

$$\{e\} = G_0 \subseteq G_1 \subseteq \ldots \subseteq G_n = G$$

of a group G is called a *normal series* of G, if G_i is a normal subgroup of G_{i+1}, for $i = 0, \ldots, n-1$.

Definition 1.8.2. A group G is said to be *solvable* if it has a normal series $\{e\} = G_0 \subseteq G_1 \subseteq \ldots \subseteq G_n = G$ such that each of its factor group G_{i+1}/G_i is abelian, for every $i = 0, \ldots, n-1$.

The above series is referred to as a *solvable series* of G.

Example 1.8.3.

(1) Abelian groups are solvable.
(2) $\mathbb{S}_3$ and $\mathbb{S}_4$ are solvable.
(3) The dihedral groups are solvable.
(4) Any group whose order has the form p^m, where p is prime, is solvable.

Theorem 1.8.4. *Any subgroup H of a solvable group G is solvable.*

Proof. Suppose that $\{e\} = G_0 \subseteq G_1 \subseteq \ldots \subseteq G_n = G$ is a solvable series for G. We show that

$$\{e\} = H \cap G_0 \subseteq H \cap G_1 \subseteq \ldots \subseteq H \cap G_n = H$$

is a solvable series for H. Since G_i is normal in G_{i+1}, where $i = 0, \ldots, n-1$, we conclude that $N_i = H \cap G_i$ is normal in $N_{i+1} = H \cap G_{i+1}$. Now, we define a mapping $f : N_{i+1} \longrightarrow G_{i+1}/G_i$ by $f(x) = xG_i$, for all $x \in N_{i+1}$. Clearly, f is a homomorphism. Moreover, if $x \in N_{i+1}$, then

$$x \in ker f \Leftrightarrow xG_i = G_i \Leftrightarrow x \in G_i \Leftrightarrow x \in H \cap G_i.$$

This yields that $ker f = H \cap G_i = N_i$. Hence, by the fundamental theorem of group homomorphism, $N_{i+1}/N_i \cong f(N_{i+1})$. As $f(N_{i+1})$ is a subgroup of G_{i+1}/G_i and G_{i+1}/G_i is abelian, $f(N_{i+1})$ is also abelian. Consequently, N_{i+1}/N_i is abelian. This proves that H is a solvable group. ∎

Theorem 1.8.5. *If N is a normal subgroup of a solvable group G, then G/N is also solvable.*

Proof. Suppose that $\{e\} = G_0 \subseteq G_1 \subseteq \ldots \subseteq G_n = G$ is a solvable series for G. We consider the following series

$$\{N\} = G_0N/N \subseteq G_1N/N \subseteq \ldots \subseteq G_nN/N = G/N.$$

Let $0 \leq i \leq n-1$ be an arbitrary element. Suppose that $x \in G_{i+1}N$. Then, $x = gy$ for some $g \in G_{i+1}$ and $y \in N$. Hence,

$$xG_iN = gyG_iN = gyNG_i = gNG_i$$
$$= G_igN = G_igyN = G_iNgy = G_iNx.$$

This proves that $G_i N$ is a normal subgroup of $G_{i+1} N$. Thus,

$$G_{i+1} N / G_i N \cong (G_{i+1} N / N) / (G_i N / N).$$

Now, we define $f : G_{i+1} \longrightarrow G_{i+1} N / G_i N$ by $f(x) = G_i N x$, for all $x \in G_{i+1}$. Then, f is a homomorphism. As $G_{i+1} N = N G_{i+1}$, for $y \in G_{i+1} N$, we can write $y = zg$ for some $z \in N$ and $g \in G_{i+1}$. Then, $G_i N y = G_i N z g = G_i N g = f(g)$. This shows that f is onto. Since $G_i \subseteq \ker f$, the following function is a homomorphism

$$\overline{f} : G_{i+1} / G_i \longrightarrow G_{i+1} N / G_i N$$
$$\overline{f}(G_i x) = G_i N x,$$

for all $x \in G_{i+1}$. Clearly, $\overline{f}$ is also onto. Thus, $G_{i+1} N / G_i N$ is a homomorphic image of the abelian group G_{i+1} / G_i. So that it must be itself abelian. Consequently, each factor group is abelian. This prove that G/N is solvable. ∎

Theorem 1.8.6. *Let N be a normal subgroup of a group G. If both N and G/N are solvable, then G is solvable.*

Proof. Since N and G/N are solvable, there exist the following solvable series for N and G/N, respectively,

$$\{e\} = N_0 \subseteq N_1 \subseteq \ldots \subseteq N_m = N,$$
$$\{N\} = G_0 / N \subseteq G_1 / N \subseteq \ldots \subseteq G_n / N = G/N.$$

Here each G_i is a subgroup of G containing N. Since G_i / N is normal in G_{i+1} / N, each G_i is normal in G_{i+1}. Moreover, $G_0 = N$. Now,

$$\{e\} = N_0 \subseteq N_1 \subseteq \ldots \subseteq N_m = N = G_0 \subseteq G_1 \subseteq \ldots \subseteq G_n = G$$

is a solvable series for G. Therefore, we conclude that G is solvable. ∎

Corollary 1.8.7. *If two groups H and K are solvable, then $H \times K$ is solvable.*

Proof. Let $G = H \times K$. Alternatively, we can consider $H \cong (H \times \{e\})$ and $K \cong (\{e\} \times H)$ and clearly, $G = (H \times \{e\})(\{e\} \times K)$. Hence, $G/H \cong K$. Since K is solvable, we have G/H is solvable. Hence, G is solvable by Theorem 1.8.6. ∎

Lemma 1.8.8. *Let N be a normal subgroup of G and G/N be abelian. If $x, y \in G$, then $xyx^{-1}y^{-1} \in N$.*

Proof. It is straightforward. ∎

Proposition 1.8.9. $\mathbb{S}_n$ *is not solvable for $n \geq 5$.*

Proof. Suppose that $\mathbb{S}_k$ is solvable for some $k \geq 5$ and

$$\{e\} = G_0 \subseteq G_1 \subseteq \ldots \subseteq G_{m-1} \subseteq G_m = \mathbb{S}_k$$

is a solvable series for $\mathbb{S}_k$. By induction, we show that G_{m-i} contains all of the cycles of length 3, where $0 \leq i \leq m$. Indeed, since $G_0 = \{e\}$, we obtain a contradiction and the proof completes.

Let $(x \ y \ z)$ be a cycle of length 3 in $\mathbb{S}_k$. Since $k \geq 5$, there exist u, v distinct from x, y, z. Now, suppose that $\sigma = (z \ u \ y)$ and $\delta = (y \ x \ v)$. Since G_{m-1} is a normal subgroup of $\mathbb{S}_k$ and $\mathbb{S}_k/G_{m-1}$ is abelian, by Lemma 1.8.8, we obtain $\sigma\delta\sigma^{-1}\delta^{-1} \in G_{m-1}$. Therefore,

$$(x \ y \ z) = (z \ u \ y)(y \ x \ v)(z \ y \ u)(y \ v \ x) = \sigma\delta\sigma^{-1}\delta^{-1}$$

is an element of G_{m-1}.

Now, let G_{m-j} contains all of the cycles of length 3. Since $(x \ y \ z) \in G_{m-j}$ and G_{m-j-1} is a normal subgroup of G_{m-j}, similar to the above discussion, we obtain $(x \ y \ z) \in G_{m-j-1}$. ∎

The *commutator subgroup* G' of a group G is the subgroup generated by the set $\{x^{-1}y^{-1}xy \mid x, y \in G\}$. That is, every element of G' has the form $a_1^{i_1} a_2^{i_2} \ldots a_k^{i_k}$, where each a_j has the form $x^{-1}y^{-1}xy$, each $i_j = \pm 1$ and k is any positive integer. $x^{-1}y^{-1}xy$ is called the *commutator* of x and y. The commutator subgroup is also called a derived subgroup.

Lemma 1.8.10. *Let G be a group and G' be its commutator subgroup. Then,*

(1) G' is a normal subgroup of G;
(2) For any normal subgroup N of G, G/N is an abelian group if and if N contains G'.

Proof. (1) Assume that $a, b \in G$. Since $(a^{-1}b^{-1}ab)^{-1} = b^{-1}a^{-1}ba$ is again a commutator, we conclude that each element of G' is a product of finite number of commutators. Now, suppose that $a \in G$ and $x \in G'$. Then, $x = g_1 g_2 \ldots g_k$, where for each $i = 1, \ldots, k$, g_i is a commutator, so $g_i = a_i^{-1} b_i^{-1} a_i b_i$ for some $a_i, b_i \in G$. So, we have $a^{-1}xa = (a^{-1}g_1 a)(a^{-1}g_2 a)\ldots(a^{-1}g_k a)$. Further,

$$a^{-1}g_i a = a^{-1}a_i^{-1}b_i^{-1}a_i b_i a$$
$$= (a^{-1}a_i a)^{-1}(a^{-1}b_i a)^{-1}(a^{-1}a_i a)(a^{-1}b_i a).$$

Hence, $a^{-1}g_i a$ is again a commutator. Therefore, $a^{-1}xa$ is a product of commutators which implies that $a^{-1}xa \in G'$.

(2) Suppose that G/N is abelian and a, b are two elements of G. Then, $(aN)(bN) = (bN)(aN)$ or $abN = baN$. So, $a^{-1}b^{-1}ab \in N$. We conclude that N contains every commutator $a^{-1}b^{-1}ab$. Since G' is generated by all the commutators, $G' \subseteq N$.

Conversely, suppose that $G' \subseteq N$ and a, b are two elements of G. Then, $a^{-1}b^{-1}ab \in G'$ gives $a^{-1}b^{-1}ab \in N$ which implies that $abN = baN$ or $(aN)(bN) = (bN)(aN)$. This shows that G/N is abelian. ∎

Corollary 1.8.11. G/G' *is abelian.*

Corollary 1.8.12. *A group G is abelian if and only if $G' = \{e\}$.*

Note that by using Corollary 1.8.12 and Theorem 1.6.11, one can give a short proof for Proposition 1.8.9.

Definition 1.8.13. Let G be a group. We define the sequence of subgroups $G^{(i)}$ of G inductively by

(1) $G^{(0)} = G$;
(2) $G^{(1)} = G'$, commutator subgroup of G;
(3) $G^{(i)} = (G^{(i-1)})'$, commutator subgroup of $G^{(i-1)}$, if $i > 1$.

Theorem 1.8.14. *A group G is solvable if and only if $G^{(n)} = \{e\}$ for some $n \geq 1$.*

Proof. Suppose that G is solvable and $\{e\} = G_0 \subseteq G_1 \subseteq \ldots \subseteq G_n = G$ is a solvable series for G. We prove inductively that
$$G^{(k)} \subseteq G_{n-k} \qquad (*)$$
for all $0 \leq k \leq n$. If $k = 0$, then $G^{(0)} = G = G_0$. Let $G^{(k)} \subseteq G_{n-k}$ for some k. This implies that $G^{(k+1)} = (G^{(k)})' \subseteq (G_{n-k})'$. Since G_{n-k}/G_{n-k-1} is abelian, by Lemma 1.8.10, $(G_{n-k})' \subseteq G_{n-k-1}$. Consequently, $G^{(k+1)} \subseteq G_{n-k-1}$. Thus, by induction, $(*)$ holds. In particular, $G^{(n)} \subseteq G_0$. Therefore, $G^{(n)} = \{e\}$.

Conversely, assume that $G^{(n)} = \{e\}$ for some n. Then
$$\{e\} = G^{(n)} \subseteq G^{(n-1)} \subseteq \ldots \subseteq G^{(1)} \subseteq G^{(0)} = G$$
is a normal series for G such that $G^{(j)}/G^{(j+1)} = G^{(j)}/(G^{(j)})'$ is abelian. Thus, G is solvable. ∎

We now introduce the class of nilpotent groups.

Definition 1.8.15. A group G is said to be *nilpotent* if it has a series of subgroups
$$\{e\} = G_0 \subseteq G_1 \subseteq \ldots \subseteq G_n = G$$
such that for every $1 \leq i \leq n$,

(1) G_i is a normal subgroup of G;

(2) $G_i/G_{i-1} \leq Z(G/G_{i-1})$.

Such series is called a *central series* of G. For a nilpotent group, the smallest n such that G has a central series of length n is called the *nilpotency class* of G; and G is said to be *nilpotent of class n*.

Let H and K be subgroups of G. We define

$$[H, K] = <h^{-1}k^{-1}hk \mid h \in H, \ k \in K>.$$

Lemma 1.8.16. *A series of G, say $\{e\} = G_0 \subseteq G_1 \subseteq \ldots \subseteq G_n = G$ is a central series if and only if for each $i = 1, \ldots, n$,*

$$[G_i, G] \leq G_{i-1}.$$

Proof. If the given series is a central series, then for each $i = 1, \ldots, n$, G_{i-1} is a normal subgroup of G and $G_i/G_{i-1} \leq Z(G/G_{i-1})$. Then, for any $x \in G_i$ and $y \in G$,

$$(xG_{i-1})(yG_{i-1}) = (yG_{i-1})(xG_{i-1}),$$

that is $xyG_{i-1} = yxG_{i-1}$. Hence, $x^{-1}y^{-1}xy \in G_{i-1}$. This implies that $[G_i, G] \leq G_{i-1}$.

Conversely, suppose that for each $i = 1, \ldots, n$, $[G_i, G] \leq G_{i-1}$. Let $x \in G_i$ and $y \in G$. Then, $x^{-1}y^{-1}xy \in G_{i-1}$. In particular, since $G_{i-1} \leq G_i$, if $x \in G_{i-1}$ then $y^{-1}xy \in G_{i-1}$. Thus, G_{i-1} is a normal subgroup of G. Moreover, $xyG_{i-1} = yxG_{i-1}$ for every $x \in G_i$ and $y \in G$, and so $G_i/G_{i-1} \leq Z(G/G_{i-1})$. Thus, the series is a central series. ∎

Example 1.8.17.

(1) Abelian groups are nilpotent.

(2) The quaternion group Q_8 is nilpotent.

Theorem 1.8.18. *If G is nilpotent, then all subgroups and all quotient groups of G are nilpotent.*

Proof. Suppose that G is nilpotent and $\{e\} = G_0 \subseteq G_1 \subseteq \ldots \subseteq G_n = G$ is a central series for G. Let H be a subgroup of G and N be a normal subgroup of G. Then, we have

$$\{e\} = H \cap G_0 \subseteq H \cap G_1 \subseteq \ldots \subseteq H \cap G_n = H, \qquad (*)$$
$$\{N\} = G_0 N/N \subseteq G_1 N/N \subseteq \ldots \subseteq G_n N/N = G/N. \qquad (**)$$

For each $i = 1, \ldots, n$, $[G_i, G] \leq G_{i-1}$. Hence,

$$[G_i \cap H, H] \leq H \cap [G_i, G] \leq H \cap G_{i-1}$$

and

$$[G_i N/N, G/N] = [G_i, G]N/N \le G_{i-1}N/N.$$

Therefore by Lemma 1.8.15, $(*)$ and $(**)$ are central series, so that H and G/N are nilpotent. ∎

In the following lemma, we present a non nilpotent group.

Lemma 1.8.19. $\mathbb{S}_3$ *is not nilpotent.*

Proof. Assume that $\mathbb{S}_3$ is nilpotent. Then, there exist a central series $\{e\} = G_0 \subseteq G_1 \subseteq \ldots \subseteq G_n = \mathbb{S}_3$. We have $G_1/G_0 \le Z(\mathbb{S}_3/G_0)$. So, $G_1 \le Z(\mathbb{S}_3) = \{e\}$ which implies that $G_1 = \{e\}$. Similarly, we obtain $G_2 = \{e\}, \ldots, G_n = \mathbb{S}_3 = \{e\}$, that is a contradiction. ∎

By the above lemma, we see that Theorem 1.8.6 is not true for nilpotent groups, i.e., if N is a normal subgroup of a group G and both N and G/N are nilpotent, then G is not nilpotent, in general.

Lemma 1.8.20. *If two groups H and K are nilpotent, then $H \times K$ is nilpotent.*

Proof. If H and K are both nilpotent, then there are central series

$$\{e\} = H_0 \subseteq H_1 \subseteq \ldots \subseteq H_m = H,$$
$$\{e\} = K_0 \subseteq K_1 \subseteq \ldots \subseteq K_n = K.$$

By inserting repetition of terms if necessary, we may assume without loss generality that $m = n$. Then, we have

$$(\{e\} \times \{e\}) = (H_0 \times K_0) \subseteq (H_1 \times K_1) \subseteq \ldots \subseteq (H_n \times K_n) = G.$$

For each $i = 1, \ldots, n$,

$$[H_i \times K_i, G] = ([H_i, H] \times [K_i, K]) \le (H_{i-1} \times K_{i-1}).$$

Now, by Lemma 1.8.16, we conclude that G is nilpotent. ∎

Definition 1.8.21. We define subgroups $\Gamma_n(G)$ and $Z_n(G)$ of G respectively as follows. Let $\Gamma_1(G) = G$ and $Z_0(G) = \{e\}$. Then, for each integer $n > 1$,

$$\Gamma_n(G) = [\Gamma_{n-1}(G), G]$$

and for each integer $n > 0$,

$$Z_n(G)/Z_{n-1}(G) = Z(G/Z_{n-1}(G)).$$

Then,

$$G = \Gamma_1(G) \geq \Gamma_2(G) \geq \Gamma_3(G) \geq \ldots,$$
$$\{e\} = Z_0(G) \leq Z_1(G) \leq Z_2(G) \leq \ldots.$$

The first sequence is called the *lower central series* of G and the second sequence is called the *upper central series* of G.

It is not difficult to see that the terms of the lower and upper central series are normal subgroups of G.

Theorem 1.8.22. *The following three statements are equivalent:*

(1) G is nilpotent.
(2) $\Gamma_n(G) = \{e\}$ for some integer n.
(3) $Z_n(G) = G$ for some integer n.

Proof. If $\Gamma_n(G) = \{e\}$ for some n, then

$$G = \Gamma_1(G) \geq \Gamma_2(G) \geq \ldots \geq \Gamma_n(G) = \{e\}$$

is a central series of G, and so G is nilpotent.

Similarly, if $Z_n(G) = G$ for some n, then

$$\{e\} = Z_0(G) \leq Z_1(G) \leq \ldots \leq Z_n(G) = G$$

is a central series of G, and so G is nilpotent.

Conversely, assume that G is nilpotent and

$$\{e\} = G_0 \subseteq G_1 \subseteq \ldots \subseteq G_n = G$$

is a central series of G. We prove first by induction on i that $G_i \leq Z_i(G)$, for all $0 \leq i \leq n$. This is trivial for $i = 0$. Suppose that $i > 0$ and, inductively, that $G_{i-1} \leq Z_{i-1}(G)$. Then, $G_{i-1}Z_{i-1}(G) = Z_{i-1}(G)$. By hypothesis, $G_i/G_{i-1} \leq Z(G/G_{i-1})$. Therefore,

$$G_i Z_{i-1}(G)/Z_{i-1}(G) \leq Z(G/Z_{i-1}(G)) = Z_i(G)/Z_{i-1}(G).$$

Hence, $G_i \leq Z_i(G)$. Thus, the induction argument goes through. In particular, since $G = G_n \leq Z_n(G)$, we obtain $Z_n(G) = G$.

Now, we prove by induction on j that

$$\Gamma_{j+1}(G) \leq G_{n-j},$$

for all $0 \leq j \leq n$. This is trivial for $j = 0$. Assume that $j > 0$ and, inductively, that $\Gamma_j(G) \leq G_{n-j+1}$. Then, by Lemma 1.8.16, $[G_{n-j+1}, G] \leq G_{n-j}$. Hence,

$$\Gamma_{j+1}(G) = [\Gamma_j(G), G] \leq [G_{n-j+1}, G] \leq G_{n-j}.$$

Again, the induction argument goes through. In particular, since $\Gamma_{n+1}(G) \le G_0 = \{e\}$, we obtain $\Gamma_{n+1}(G) = \{e\}$. ∎

Corollary 1.8.23. *Let G be a nilpotent group. Then, for any central series of G, say $\{e\} = G_0 \subseteq G_1 \subseteq \ldots \subseteq G_n = G$, we have*

$$\Gamma_{n-i+1}(G) \le G_i \le Z_i(G),$$

for all $0 \le i \le n$. Moreover, the least integer r such that $\Gamma_{r+1}(G) = \{e\}$ is equal to the least integer r such that $Z_r(G) = G$.

Proof. Suppose that r is the least integer such that $Z_r(G) = G$. We show that r is also the least integer such that $\Gamma_{r+1}(G) = \{e\}$. Let $\Gamma_{k+1}(G) = \{e\}$ for some $k < r$. We set $G_i = \Gamma_{k-i+1}(G)$. Then,

$$G_k = \Gamma_1(G), \ldots, G_0 = \Gamma_{k+1}(G) = \{e\}.$$

Hence, $\{e\} = G_0 \subseteq G_1 \subseteq \ldots \subseteq G_k = G$ is a central series. Now, by Theorem 1.8.22, it follows that $Z_k(G) = G$; this is contrary to the definition of r. ∎

We conclude that if G is a nilpotent group, then the lower central series of G is its most rapidly descending central series and the upper central series of G is its most rapidly ascending central series.

Lemma 1.8.24. *For each non-negative integer i, $G^{(i)} \le \Gamma_{i+1}(G)$.*

Proof. We prove by induction on i. If $i = 0$, then $G^{(0)} = G = \Gamma_1(G)$. Suppose that $i > 0$ and, inductively, that $G^{(i-1)} \le \Gamma_i(G)$. Then,

$$G^{(i)} = [G^{(i-1)}, G^{(i-1)}] \le [\Gamma_i(G), G] = \Gamma_{i+1}(G).$$

This completes the proof. ∎

Theorem 1.8.25. *Each nilpotent group is solvable.*

Proof. Suppose that G is a nilpotent group. By Corollary 1.8.23, there exist a non-negative integer r such that $\Gamma_{r+1} = \{e\}$. By Lemma 1.8.24, we get $G^{(i)} \le \Gamma_{i+1}(G) = \{e\}$. Now, by Theorem 1.8.14, we conclude that G is solvable. ∎

Lemma 1.8.19 shows that the converse of Theorem 1.8.25 is not true in general.

Chapter 2

Hypergroups

2.1 Introduction and historical development of hypergroups

The hypergroup notion was introduced in 1934 by F. Marty [103], at the 8th Congress of Scandinavian Mathematicians. He published some notes on hypergroups, using them in different contexts: algebraic functions, rational fractions, non-commutative groups. Hypergroups are a suitable generalization of groups. We know in a group, the composition of two elements is an element, while in a hypergroup, the composition of two elements is a set. The motivating example was the following: Let G be a group and K be a subgroup of G. Then, $G/K = \{xK \mid x \in G\}$ becomes a hypergroup, where the composition is defined in a usual manner.

In [119], Prenowitz represented several kinds of geometries (projective, descriptive and spherical) as hypergroups, and later, with Jantosciak [124], founded geometries on join spaces, a special kind of hypergroups, which in the last decades were shown to be useful instrument in the study of several matters: graphs, hypergraphs and binary relations.

Several kinds of hypergroups have been intensively studied, such as regular hypergroups, reversible regular hypergroups, canonical hypergroups, cogroups and cyclic hypergroups. The situations that occur in hypergroup theory, are often extremely diversified and complex with respect to group theory. For instance, there are homomorphisms of various types between hypergroups and there are several kinds of subhypergroups, such as: closed, invertible, ultraclosed, conjugable.

Around the 1940's, the general aspects of the theory, the connections with groups and various applications in geometry were studied in France by F. Marty, M. Krasner, M. Kuntzmann, R. Croisot, in U.S.A. by M. Dresher,

O. Ore, W. Prenowitz, H.S. Wall, J.E. Eaton, H. Campaigne, L. Griffiths, in Russia by A. Dietzman, A. Vikhrov, in Italy by G. Zappa, in Japan by Y. Utumi.

Over the following two decades, other interesting results on hyperstructures were obtained, for instance, in Italy, A. Orsatti studied semiregular hypergroups, in Czechoslovakia, K. Drbohlav studied hypergroups of two sided classes, in Romania, M. Benado studied hyperlattices.

The theory knew an important progress starting with the 1970's, when its research area enlarged. In France, M. Krasner, M. Koskas and Y. Sureau investigated the theory of subhypergroups and the relations defined on hyperstructures; in Greece, J. Mittas, and his students M. Konstantinidou, K. Serafimidis, S. Ioulidis and C.N. Yatras studied the canonical hypergroups, the hyperrings, the hyperlattices, Ch. Massouros obtained important results about hyperfields and other hyperstructures. G. Massouros, together with J. Mittas studied applications of hyperstructures to Automata. D. Stratigopoulos continued some of Krasner ideas, studying in depth noncommutative hyperrings and hypermodules. T. Vougiouklis, L. Konguetsof and later S. Spartalis, A. Dramalidis analyzed especially the cyclic hypergroups, the P-hyperstructures.

Significant contributions to the study of regular hypergroups, complete hypergroups, of the heart and of the hypergroup homomorphisms in general or with applications in Combinatorics and Geometry were brought by the Italian mathematician P. Corsini and his group of research, among whom we mention M.de Salvo, R. Migliorato, F. de Maria, G. Romeo, P. Bonansinga.

Also around 1970's, some connections between hyperstructures and ordered systems, particularly lattices, were established by T. Nakano and J.C. Varlet. Around the 1980's and 90's, associativity semyhypergroups were analyzed in the context of semigroup theory by T. Kepka and then by J. Jezec, P. Nemec and K. Drbohlav, and in Finland by M. Niemenmaa.

In U.S.A., R. Roth used canonical hypergroups in solving some problems of character theory of finite groups, while S. Comer studied the connections among hypergroups, combinatorics and the relation theory. J. Jantosciak continued the study of join spaces, introduced by W. Prenowitz, he considered a generalization of them for the noncommutative case and studied correspondences between homomorphisms and the associated relations.

In America, hyperstructures have been studied both in U.S.A. (at Charleston, South Carolina – The Citadel, New York-Brooklyn College, CUNY, Cleveland, Ohio – John Carroll University) and in Canada (at Université de Montréal).

A big role in spreading this theory is played by the Congresses on Algebraic Hyperstructures and their Applications.

The first three Congresses were organized by P. Corsini in Italy. The contribution of P. Corsini in the development of Hyperstructure Theory has been decisive. He has delivered lectures about hyperstructures and their applications in several countries, several times, for instance in Romania, Thailand, Iran, China, Montenegro, making known this theory. After his visits in these countries, hyperstructures have had a substantial development.

Coming back to the Congresses on Algebraic Hyperstructures, the first two were organized in Taormina, Sicily, in 1978 and 1983, with the names: "Sistemi Binari e loro Applicazioni" and "Ipergruppi, Strutture Multivoche e Algebrizzazione di Strutture d'Incidenza". The third Congress, called "Ipergruppi, altre Strutture Multivoche e loro Applicazioni" was organized in Udine in 1985.

The fourth congress, organized by T. Vougiouklis in Xanthi in 1990, used already the name of Algebraic Congress on Hyperstructures and their Applications, also known as AHA Congress. After 1990, AHA Congresses have been organized every three years. Beginning with the 90's Hyperstructure Theory represents a constant concern also for the Romanian mathematicians, the decisive moment being the fifth AHA Congress, organized in 1993 at the University "Al.I.Cuza" of Iasi by M. Stefanescu. This domain of the modern algebra is a topic of a great interest also for the Romanian researches, who have published a lot of papers on hyperstructures in national or international journals, have given communications in conferences and congresses, have written Ph.D. theses in this field.

The sixth AHA Congress was organized in 1996 at the Agriculture University of Prague by T. Kepka and P. Nemec, the seventh was organized in 1999 by R. Migliorato in Taormina, Sicily, then the eighth was organized in 2002 by T. Vougiouklis in Samothraki, Greece. All these congresses were organized in Europe. Nowadays, one works successfully on Hyperstructures in the following countries of Europe:

- in Greece, at Thessaloniki (Aristotle University), at Alexandropoulis (Democritus University of Thrace), at Patras (Patras University), Orestiada (Democritus University of Thrace), at Athens;
- in Italy at Udine University, at Messina University, at Rome (Uni-

versita' "La Sapienza"), at Pescara (D'Annunzio University), at
Teramo (Universita' di Teramo), Palermo University;
- in Romania, at Iasi ("Al.I. Cuza" University), Cluj ("Babes-
Bolyai" University), Constanta ("Ovidius" University);
- in Czech Republic, at Praha (Charles University, Agriculture
University), at Brno (Brno University of Technology, Military
Academy of Brno, Masaryk University), Olomouc (Palacky University);
- in Montenegro, at Podgorica University.

Let us continue with the following AHA Congresses.

The ninth congress on hyperstructures, organized in 2005 by R. Ameri
in Babolsar, Iran, was the first of this kind in Asia. In the past millenniums,
Iran gave fundamental contributions to Mathematics and in particular, to
Algebra (for instance Khwarizmi, Kashi, Khayyam and recently Zadeh),
many scientists have well understood the importance of hyperstructures,
on the theoretical point of view and for the applications to a wide variety
of scientific sectors.

Nowadays, hyperstructures are cultivated in many universities and research centers in Iran, among which we mention Yazd University, Shahid
Bahonar University of Kerman, Mazandaran University, Kashan University, Ferdowsi University of Mashhad, Tehran University, Tarbiat Modarres
University, Zahedan (Sistan and Baluchestan University), Semnan University, Islasmic Azad University of Kerman, Shahid Beheshti University of
Tehran, Center for Theoretical Physics and Mathematics of Tehran, Zan-
jan (Institute for Advanced Studies in Basic Sciences).

Another Asian country where hyperstructures have had success is Thailand. In Chulalornkorn University of Bangkok, important results have been
obtained by Y. Kemprasit and her students Y. Punkla, S. Chaoprakhoi,
N. Triphop, C. Namnak on the connections among hyperstructures, semigroups and rings.

There are other Asia centers for researches in hyperstructures. We mention here India (University of Calcutta, Aditanar College of Arts and Sciences, Tiruchendur, Tamil Nadu), Korea (Chiungju National University,
Chiungju National University of Education, Gyeongsang National University, Jinju), Japan (Hitotsubashi University of Tokyo), Sultanate of Oman
(Education College for Teachers), China (Northwest University of Xian,
Yunnan University of Kunming).

Hypertructures have been also cultivated in Germany, Netherlands, Bel-
gium, Macedonia, Serbia, Slovakia, Spain, Uzbekistan, Australia. The

tenth AHA Congress was held in Brno, Czech Republic in the autumn of 2008. It was organized by Šárka. Hošková, at the Military Academy of Brno. The eleventh AHA Congress was held in Pescara, Italy in the autumn of 2011. It was organized by A. Maturo, at the Università degli Studi "G. d'Annunzio" Chieti-Pescara.

More than 800 papers and some books have been written till now on hyperstructures. Many of them are dedicated to the applications of hyperstructures in other topics. We shall mention here some of the fields connected with hyperstructures and only some names of mathematicians who have worked in each topic:

- *Geometry* (W. Prenowitz, J. Jantosciak, and later G. Tallini),
- *Codes* (G. Tallini),
- *Cryptography and Probability* (L. Berardi, F. Eugeni, S. Innamorati, A. Maturo),
- *Automata* (G. Massouros, J. Chvalina, L. Chvalinova),
- *Artificial Intelligence* (G. Ligozat),
- *Median Algebras, Relation Algebras*, C-algebras (S. Comer),
- *Boolean Algebras* (A.R. Ashrafi, M. Konstantinidou),
- *Categories* (M. Scafati, M.M. Zahedi, C. Pelea, R. Bayon, N. Ligeros, S.N. Hosseini, B. Davvaz, M.R. Khosharadi-Zadeh),
- *Topology* (J. Mittas, M. Konstantinidou, M.M. Zahedi, R. Ameri, S. Hošková),
- *Binary Relations* (J. Chvalina, I.G. Rosenberg, P. Corsini, V. Leoreanu, B. Davvaz, S. Spartalis, I. Chajda, S. Hošková, I. Cristea, M. De Salvo, G. Lo Faro),
- *Graphs and Hypergraphs* (P. Corsini, I.G. Rosenberg, V. Leoreanu, M. Gionfriddo, A. Iranmanesh, M.R. Khosharadi-Zadeh),
- *Lattices and Hyperlattices* (J.C. Varlet, T. Nakano, J. Mittas, A. Kehagias, M. Konstantinidou, K. Serafimidis, V. Leoreanu, I.G. Rosenberg, B. Davvaz, G. Calugareanu, G. Radu, A.R. Ashrafi),
- *Fuzzy Sets and Rough Sets* (P. Corsini, M.M. Zahedi, B. Davvaz, R. Ameri, R.A. Borzooei, V. Leoreanu, I. Cristea, A. Kehagias, A. Hasankhani, I. Tofan, C. Volf, G.A. Moghani, H. Hedayati),
- *Intuitionistic Fuzzy Hyperalgebras* (B. Davvaz, R.A. Borzooei, Y.B. Jun, W.A. Dudek, L. Torkzadeh),
- *Generalized Dynamical Systems* (M.R. Molaei) and so on.

Another topic which has aroused the interest of several mathematicians, is that one of H_v-structures, introduced by T. Vougiouklis and studied then also by B. Davvaz, M.R. Darafsheh, M. Ghadiri, R. Migliorato, S. Spartalis, A. Dramalidis, A. Iranmanesh, M.N. Iradmusa, A. Madanshekaf. H_v-structures are a special kind of hyperstructures, for which the weak associativity holds. Recently, n-ary hyperstructures, introduced by B. Davvaz and T. Vougiouklis, represent an intensively studied field of research.

Therefore, there are good reasons to hope that Hyperstructure Theory will be one of the more successful fields of research in algebra.

2.2   Definition and examples of hypergroups

Let H be a non-empty set and $\circ : H \times H \longrightarrow \mathcal{P}^*(H)$ be a *hyperoperation*, where $\mathcal{P}^*(H)$ is the family of non-empty subsets of H. The couple $(H, \circ)$ is called a *hypergroupoid*. For any two non-empty subsets A and B of H and $x \in H$, we define

$$A \circ B = \bigcup_{a \in A, b \in B} a \circ b, \quad A \circ x = A \circ \{x\} \quad \text{and} \quad x \circ B = \{x\} \circ B.$$

Definition 2.2.1. A hypergroupoid $(H, \circ)$ is called a *semihypergroup* if for all a, b, c of H we have $(a \circ b) \circ c = a \circ (b \circ c)$, which means that

$$\bigcup_{u \in a \circ b} u \circ c = \bigcup_{v \in b \circ c} a \circ v.$$

A hypergroupoid $(H, \circ)$ is called a *quasihypergroup* if for all a of H we have $a \circ H = H \circ a = H$. This condition is also called the *reproduction axiom*.

Definition 2.2.2. A hypergroupoid $(H, \circ)$ which is both a semihypergroup and a quasihypergroup is called a *hypergroup*.

A hypergroup for which the hyperproduct of any two elements has exactly one element is a group. Indeed, let $(H, \circ)$ be a hypergroup, such that for all x, y of H, we have $|x \circ y| = 1$. Then, $(H, \circ)$ is a semigroup, such that for all a, b in H, there exist x and y for which we have $a = b \circ x$ and $a = y \circ b$. It follows that $(H, \circ)$ is a group.

Now, we look at some examples of hypergroups.

Example 2.2.3.

(1) If H is a non-empty set and for all x, y of H, we define $x \circ y = H$, then $(H, \circ)$ is a hypergroup, called the *total hypergroup*.

(2) Let $(S, \cdot)$ be a semigroup and P be a non-empty subset of S. For all x, y of S, we define $x \circ y = xPy$. Then, $(S, \circ)$ is a semihypergroup. If $(S, \cdot)$ is a group, then $(S, \circ)$ is a hypergroup, called a *P-hypergroup*.

(3) If G is a group and for all x, y of G, $< x, y >$ denotes the subgroup generated by x and y, then we define $x \circ y = < x, y >$. We obtain that $(G, \circ)$ is a hypergroup.

(4) If $(G, \cdot)$ is a group, H is a normal subgroup of G and for all x, y of G, we define $x \circ y = xyH$, then $(G, \circ)$ is a hypergroup.

(5) Let $(G, \cdot)$ be a group and let H be a non-normal subgroup of it. If we denote $G/H = \{xH \mid x \in G\}$, then $(G/H, \circ)$ is a hypergroup, where for all xH, yH of G/H, we have $xH \circ yH = \{zH \mid z \in xHy\}$.

(6) If $(G, +)$ is an abelian group, ρ is an equivalence relation in G, which has classes $\overline{x} = \{x, -x\}$, then for all $\overline{x}, \overline{y}$ of G/ρ, we define $\overline{x} \circ \overline{y} = \{\overline{x + y}, \overline{x - y}\}$. We obtain that $(G/\rho, \circ)$ is a hypergroup.

(7) Let D be an integral domain and let F be its field of fractions. If we denote by U the group of the invertible elements of D, then we define the following hyperoperation on F/U: for all $\overline{x}, \overline{y}$ of F/U, we have $\overline{x} \circ \overline{y} = \{\overline{z} \mid \exists(u, v) \in U^2 \text{ such that } z = ux + vy\}$. We obtain that $(F/U, \circ)$ is a hypergroup.

(8) Let $(L, \wedge, \vee)$ be a lattice with a minimum element 0. If for all $a \in L$, $F(a)$ denotes the principal filter generated from a, then we obtain a hypergroup $(L, \circ)$, where for all a, b of L, we have $a \circ b = F(a \wedge b)$.

(9) Let $(L, \wedge, \vee)$ be a modular lattice. If for all x, y of L, we define $x \circ y = \{z \in L \mid z \vee x = x \vee y = y \vee z\}$, then $(L, \circ)$ is a hypergroup.

(10) Let $(L, \wedge, \vee)$ be a distributive lattice. If for all x, y of L, we define $x \circ y = \{z \in L \mid x \wedge y \leq z \leq x \vee y\}$, then $(L, \circ)$ is a hypergroup.

(11) Let H be a non-empty set and $\mu : H \longrightarrow [0, 1]$ be a function. If for all x, y of H we define $x \circ y = \{z \in L \mid \mu(x) \wedge \mu(y) \leq \mu(z) \leq \mu(x) \vee \mu(y)\}$, then $(H, \circ)$ is a hypergroup.

(12) Let H be a non-empty set and R be an equivalence relation in H, such that for all x of H, the equivalence class $R(x)$ of x has at least three elements. For any subset A of H, $\overline{R}(A)$ denotes the set $\bigcup\limits_{R(x) \cap A \neq \emptyset} R(x)$, while $\underline{R}(A)$ denotes the set $\bigcup\limits_{R(x) \subseteq A} R(x)$. The couple $(\underline{R}(A), \overline{R}(A))$ is called a *rough set*. If for all x, y of H, we define $x \circ y = \overline{R}(\{x, y\}) \backslash \underline{R}(\{x, y\})$, then $(H, \circ)$ is a hypergroup.

(13) Let H be a non-empty set and μ, λ be two functions from H to $[0, 1]$. For all x, y of H we define

$$x \circ y = \{u \in H \mid \mu(x) \wedge \lambda(x) \wedge \mu(y) \wedge \lambda(y) \leq \mu(u) \wedge \lambda(u)$$
$$\text{and } \mu(u) \vee \lambda(u) \leq \mu(x) \vee \lambda(x) \vee \mu(y) \vee \lambda(y)\}.$$

Then, the hyperstructure $(H, \circ)$ is a commutative hypergroup.

(14) Define the following hyperoperation on the real set $\mathbb{R}$: for all $x \in \mathbb{R}$, $x \circ x = x$ and for all different real elements x, y, $x \circ y$ is the open interval between x and y. Then, $(\mathbb{R}, \circ)$ is a hypergroup.

2.3   Some kinds of subhypergroups

A non-empty subset K of a semihypergroup $(H, \circ)$ is called a *subsemihypergroup* if it is a semihypergroup. In other words, a non-empty subset K of a semihypergroup $(H, \circ)$ is a subsemihypergroup if $K \circ K \subseteq K$.

Definition 2.3.1. A non-empty subset K of a hypergroup $(H, \circ)$ is called a *subhypergroup* if it is a hypergroup.

Hence, a non-empty subset K of a hypergroup $(H, \circ)$ is a subhypergroup if for all a of K we have $a \circ K = K \circ a = K$.

There are several kinds of subhypergroups. In what follows, we introduce closed, invertible, ultraclosed and conjugable subhypergroups and some connections among them.

Among the mathematicians who studied this topic, we mention F. Marty, M. Dresher, O. Ore, M. Krasner who analyzed closed and invertible subhypergroups. M. Koskas considered another type of subhypergroups are complete parts. Later Y. Sureau has studied ultraclosed, invertible and conjugable subhypergroups. Corsini has obtained important results about ultraclosed and complete parts. Also, Leoreanu has studied and obtained other interesting results on subhypergroups.

Let us present now the definition of these types of subhypergroups. Let $(H, \circ)$ be a hypergroup and $(K, \circ)$ be a subhypergroup of it.

Definition 2.3.2. We say that K is:

- *closed on the left (on the right)* if for all k_1, k_2 of K and x of H, from $k_1 \in x \circ k_2$ ($k_1 \in k_2 \circ x$, respectively), it follows that $x \in K$;
- *invertible on the left (on the right)* if for all x, y of H, from $x \in K \circ y$ ($x \in y \circ K$), it follows that $y \in K \circ x$ ($y \in x \circ K$, respectively);
- *ultraclosed on the left (on the right)* if for all x of H, we have $K \circ x \cap (H \backslash K) \circ x = \emptyset$ ($x \circ K \cap x \circ (H \backslash K) = \emptyset$);
- *conjugable on the right* if it is closed on the right and for all $x \in H$, there exists $x' \in H$ such that $x' \circ x \subseteq K$. Similarly, we can define the notion of *conjugable on the left*.

We say that K is *closed* (*invertible, ultraclosed, conjugable*) if it is closed (invertible, ultraclosed, conjugable, respectively) on the left and on the right.

Example 2.3.3.

(1) Let $(A, \circ)$ be a hypergroup, $H = A \cup T$, where T is a set with at least three elements and $A \cap T = \emptyset$. We define the hyperoperation $\otimes$ on H, as follows:

if $(x, y) \in A^2$, then $x \otimes y = x \circ y$;

if $(x, t) \in A \times T$, then $x \otimes t = t \otimes x = t$;

if $(t_1, t_2) \in T \times T$, then $t_1 \otimes t_2 = t_2 \otimes t_1 = A \cup (T \setminus \{t_1, t_2\})$.

Then, $(H, \otimes)$ is a hypergroup and $(A, \otimes)$ is an ultraclosed, non-conjugable subhypergroup of H.

(2) Let $(A, \circ)$ be a total hypergroup, with at least two elements and let $T = \{t_i\}_{i \in \mathbb{N}}$ such that $A \cap T = \emptyset$ and $t_i \neq t_j$ for $i \neq j$. We define the hyperoperation $\otimes$ on $H = A \cup T$ as follows:

if $(x, y) \in A^2$, then $x \otimes y = A$;

if $(x, t) \in A \times T$, then $x \otimes t = t \otimes x = (A \setminus \{x\}) \cup T$;

if $(t_i, t_j) \in T \times T$, then $t_i \otimes t_j = t_j \otimes t_i = A \cup \{t_{i+j}\}$.

Then, $(H, \otimes)$ is a hypergroup and $(A, \otimes)$ is a non-closed subhypergroup of H.

(3) Let us consider the group $(\mathbb{Z}, +)$ and the subgroups $S_i = 2^i \mathbb{Z}$, where i is a non-negative integer. For any $x \in \mathbb{Z} \setminus \{0\}$, there exists a unique integer $n(x)$, such that $x \in S_{n(x)} \setminus S_{n(x)+1}$. Define the following commutative hyperoperation on $\mathbb{Z} \setminus \{0\}$:

if $n(x) < n(y)$, then $x \circ y = x + S_{n(y)}$;

if $n(x) = n(y)$, then $x \circ y = S_{n(x)} \setminus \{0\}$;

if $n(x) > n(y)$, then $x \circ y = y + S_{n(x)}$.

Notice that if $n(x) < n(y)$, then $n(x+y) = n(x)$. Then, $(\mathbb{Z} \setminus \{0\}, \circ)$ is a hypergroup and for all $i \in \mathbb{N}$, $(S_i \setminus \{0\}, \circ)$ is an invertible subhypergroup of $\mathbb{Z} \setminus \{0\}$.

Other examples can be found in [22].

Lemma 2.3.4. *A subhypergroup K is invertible on the right if and only if $\{x \circ K\}_{x \in H}$ is a partition of H.*

Proof. If K is invertible on the right and $z \in x \circ K \cap y \circ K$, then $x, y \in z \circ K$, whence $x \circ K \subseteq z \circ K$ and $y \circ K \subseteq z \circ K$. It follows that $x \circ K = z \circ K = y \circ K$. Conversely, if $\{x \circ K\}_{x \in H}$ is a partition of H and $x \in y \circ K$, then $x \circ K \subseteq y \circ K$, whence $x \circ K = y \circ K$ and so we have $x \in y \circ K = x \circ K$. Hence, for

all x of H we have $x \in x \circ K$. From here, we obtain that $y \in y \circ K = x \circ K$.
■

Similar to Lemma 2.3.4, we can give a necessary and sufficient condition for invertible subhypergroups on the left.

The following theorems present some connections among the above types of subhypergroups.

If A and B are subsets of H such that we have $H = A \cup B$ and $A \cap B = \emptyset$, then we denote $H = A \oplus B$.

Theorem 2.3.5. *If a subhypergroup K of a hypergroup $(H, \circ)$ is ultraclosed, then it is closed and invertible.*

Proof. First we check that K is closed. For $x \in K$, we have $K \cap x \circ (H \backslash K) = \emptyset$ and from $H = x \circ K \cup x \circ (H \backslash K)$, we obtain $x \circ (H \backslash K) = H \backslash K$, which means that $K \circ (H \backslash K) = H \backslash K$. Similarly, we obtain $(H \backslash K) \circ K = H \backslash K$, hence K is closed. Now, we show that $\{x \circ K\}_{x \in H}$ is a partition of H. Let $y \in x \circ K \cap z \circ K$. It follows that $y \circ K \subseteq x \circ K$ and $y \circ (H \backslash K) \subseteq x \circ K \circ (H \backslash K) = x \circ (H \backslash K)$. From $H = x \circ K \oplus x \circ (H \backslash K) = y \circ K \oplus y \circ (H \backslash K)$, we obtain $x \circ K = y \circ K$. Similarly, we have $z \circ K = y \circ K$. Hence, $\{x \circ K\}_{x \in H}$ is a partition of H, and according to the above lemma, it follows that K is invertible on the right. Similarly, we can show that K is invertible on the left. ■

Theorem 2.3.6. *If a subhypergroup K of a hypergroup $(H, \circ)$ is invertible, then it is closed.*

Proof. Let $k_1, k_2 \in K$. If $k_1 \in x \circ k_2 \subseteq x \circ K$, then $x \in k_1 \circ K \subseteq K$. Similarly, from $k_1 \in k_2 \circ x$, we obtain $x \in K$. ■

We denote the set $\{e \in H \mid \exists x \in H, \text{ such that } x \in x \circ e \cup e \circ x\}$ by I_p and we call it the *set of partial identities* of H.

Theorem 2.3.7. *A subhypergroup K of a hypergroup $(H, \circ)$ is ultraclosed if and only if K is closed and $I_p \subseteq K$.*

Proof. Suppose that K is closed and $I_p \subseteq K$. First, we show that K is invertible on the left. Suppose there are x, y of H such that $x \in K \circ y$ and $y \notin K \circ x$. Hence, $y \in (H \backslash K) \circ x$, whence $x \in K \circ (H \backslash K) \circ x \subseteq (H \backslash K) \circ x$, since K is closed. We obtain that $I_p \cap (H \backslash K) \neq \emptyset$, which is a contradiction. Hence, K is invertible on the left. Now, we check that K is ultraclosed on the left. Suppose that there are a and x in H such that $a \in K \circ x \cap (H \backslash K) \circ x$. It follows that $x \in K \circ a$, since K is invertible on the left. We obtain

$a \in (H \setminus K) \circ x \subseteq (H \setminus K) \circ K \circ a \subseteq (H \setminus K) \circ a$, since K is closed. This means that $I_p \cap (H \setminus K) \neq \emptyset$, which is a contradiction. Therefore, K is ultraclosed on the left and similarly it is ultraclosed on the right.

Conversely, suppose that K is ultraclosed. According to Theorem 2.3.5, K is closed. Now, suppose that $I_p \cap (H \setminus K) \neq \emptyset$, which means that there is $e \in H \setminus K$ and there is $x \in H$, such that $x \in e \circ x$, for instance. We obtain $x \in (H \setminus K) \circ x$, whence $K \circ x \subseteq (H \setminus K) \circ x$, which contradicts that K is ultraclosed. Hence, $I_p \subseteq K$. ∎

Theorem 2.3.8. *If a subhypergroup K of a hypergroup $(H, \circ)$ is conjugable, then it is ultraclosed.*

Proof. Let $x \in H$. Denote $B = x \circ K \cap x \circ (H \setminus K)$. Since K is conjugable it follows that K is closed and there exists $x' \in H$, such that $x' \circ x \subseteq K$. We obtain

$$
\begin{aligned}
x' \circ B &= x' \circ (x \circ K \cap x \circ (H \setminus K)) \\
&\subseteq K \cap x' \circ x \circ (H \setminus K) \\
&\subseteq K \cap K \circ (H \setminus K) \\
&\subseteq K \cap (H \setminus K) = \emptyset.
\end{aligned}
$$

Hence, $B = \emptyset$, which means that K is ultraclosed on the right. Similarly, we check that K is ultraclosed on the left. ∎

2.4 Homomorphisms of hypergroups

Homomorphisms of hypergroups are studied by Dresher, Ore, Krasner, Kuntzmann, Koskas, Jantosciak, Corsini, Freni, Davvaz and many others. In this section, we study several kinds of homomorphisms. The main references are [22; 86].

Definition 2.4.1. Let $(H_1, \circ)$ and $(H_2, *)$ be two hypergroups. A map $f : H_1 \longrightarrow H_2$, is called

(1) a *homomorphism* or *inclusion homomorphism* if for all x, y of H_1, we have $f(x \circ y) \subseteq f(x) * f(y)$;

(2) a *good homomorphism* if for all x, y of H_1, we have $f(x \circ y) = f(x) * f(y)$;

(3) an *isomorphism* if it is a one to one and onto good homomorphism. If f is an isomorphism, then H_1 and H_2 are said to be *isomorphic*, which is denoted by $H_1 \cong H_2$.

Example 2.4.2. Let $H_1 = \{a, b, c\}$ and $H_2 = \{0, 1, 2\}$ be two hypergroups with the following hyperoperations:

$\circ$	a	b	c
a	a	H_1	H_1
b	H_1	b	b
c	H_1	b	c

$*$	0	1	2
0	0	H_2	H_2
1	H_2	1	1
2	H_2	1	$\{1, 2\}$

and let $f : H_1 \longrightarrow H_2$ is defined by $f(a) = 0$, $f(b) = 1$ and $f(c) = 2$. Clearly, f is an inclusion homomorphism.

Proposition 2.4.3. *Let* $(H_1, \circ)$ *and* $(H_2, *)$ *be two hypergroups and* $f : H_1 \longrightarrow H_2$ *be a good homomorphism. Then,* Imf *is a subhypergroup of* H_2.

Proof. For every $a, b \in H_1$, we have $f(a) * f(b) = f(a \circ b) \subseteq Imf$. Moreover, there exist $x, y \in H_1$ such that $a \in x \circ b$ and $a \in b \circ y$ and consequently $f(a) \in f(x) * f(b)$ and $f(a) \in f(b) * f(y)$. ∎

We employ for simplicity of notation $x_f = f^{-1}(f(x))$ and for a subset A of H_1, $A_f = f^{-1}(f(A)) = \cup\{x_f \mid x \in A\}$.

Notice that the defining condition for an inclusion homomorphism is equivalent to

$$x \circ y \subseteq f^{-1}(f(x) * f(y)).$$

It is also clear for an inclusion homomorphism that

$$(x \circ y)_f \subseteq f^{-1}(f(x) * f(y)).$$

The defining condition for an inclusion homomorphism is also valid for sets. That is, if A, B are non-empty subsets of H_1, then it follows that

$$f(A \circ B) \subseteq f(A) * f(B).$$

Applying the above relation for $A = x_f$ and $B = y_f$, we obtain

$$x_f \circ y_f \subseteq f^{-1}(f(x) * f(y))$$

and

$$(x_f \circ y_f)_f \subseteq f^{-1}(f(x) * f(y)).$$

Homomorphisms having various types of properties are defined and studied in the literature. Each of these properties can be viewed as a condition on $f^{-1}(f(x) * f(y))$. We consider four types of homomorphisms in the following definition.

Definition 2.4.4. Let $(H_1, \circ)$ and $(H_2, *)$ be two hypergroups and $f : H_1 \longrightarrow H_2$ be a mapping. Then, given $x, y \in H_1$, f is called a homomorphism of

type 1, if $f^{-1}(f(x) * f(y)) = (x_f \circ y_f)_f;$
type 2, if $f^{-1}(f(x) * f(y)) = (x \circ y)_f;$
type 3, if $f^{-1}(f(x) * f(y)) = x_f \circ y_f;$
type 4, if $f^{-1}(f(x) * f(y)) = (x \circ y)_f = x_f \circ y_f.$

Note that $x \circ y \subseteq (x \circ y)_f$, $x \circ y \subseteq x_f \circ y_f$ and that $(x \circ y)_f \subseteq (x_f \circ y_f)_f$, $x_f \circ y_f \subseteq (x_f \circ y_f)_f$. Hence, a homomorphism of any type 1 through 4 is indeed an inclusion homomorphism. Observe that a one to one homomorphism of H_1 onto H_2 of any type 1 through 4 is an isomorphism.

Proposition 2.4.5. *Let* $(H_1, \circ)$ *and* $(H_2, *)$ *be two hypergroups,* A, B *are non-empty subsets of* H_1 *and* $f : H_1 \longrightarrow H_2$ *be a mapping. Then,* f *is a homomorphism of*

(1) type 1 implies $f^{-1}(f(A) * f(B)) = (A_f \circ B_f)_f;$
(2) type 2 implies $f^{-1}(f(A) * f(B)) = (A \circ B)_f;$
(3) type 3 implies $f^{-1}(f(A) * f(B)) = A_f \circ B_f;$
(4) type 4 implies $f^{-1}(f(A) * f(B)) = (A \circ B)_f = A_f \circ B_f.$

Proof. Each part is established by a straightforward set theoretic argument. ■

Proposition 2.4.6. *Let* $(H_1, \circ)$ *and* $(H_2, *)$ *be two hypergroups and* $f : H_1 \longrightarrow H_2$ *be a mapping. Then,* f *is a homomorphism of*

(1) type 4 if and only if f *is a homomorphism of type 2 and type 3;*
(2) type 1 if f *is a homomorphism of type 2 or type 3.*

Proof. (1) It is trivial.
(2) Suppose that $x, y \in H_1$ and f is a homomorphism of type 2. Then,

$$(x \circ y)_f \subseteq (x_f \circ y_f)_f \subseteq f^{-1}(f(x) * f(y)) = (x \circ y)_f.$$

Similarly, if f is a homomorphism of type 3, then

$$x_f \circ y_f \subseteq (x_f \circ y_f)_f \subseteq f^{-1}(f(x) * f(y)) = x_f \circ y_f.$$

Hence, in either case, f is a homomorphism of type 1. Thus, (2) holds. ■

Homomorphism Types

The defining condition for a homomorphism of type 1 or type 2 can easily be simplified if the homomorphism is onto.

Proposition 2.4.7. *Let $(H_1, \circ)$ and $(H_2, *)$ be two hypergroups and $f : H_1 \longrightarrow H_2$ be an onto mapping. Then, given $x, y \in H_1$, f is a homomorphism of*

(1) type 1 if and only if $f(x_f \circ y_f) = f(x) * f(y)$;
(2) type 2 if and only if $f(x \circ y) = f(x) * f(y)$.

Proof. It is straightforward. ∎

Corollary 2.4.8. *Let $(H_1, \circ)$ and $(H_2, *)$ be two hypergroups, A, B be non-empty subsets of H_1 and $f : H_1 \longrightarrow H_2$ be an onto mapping. Then, f is a homomorphism of*

(1) type 1 implies $f(A_f \circ B_f) = f(A) * f(B)$;
(2) type 2 implies $f(A \circ B) = f(A) * f(B)$.

Example 2.4.9. Consider $H_1 = \{0, 1, 2\}$ and $H_2 = \{a, b\}$ together with the following hyperoperations:

$\circ$	0	1	2
0	0	$\{0,1\}$	$\{0,2\}$
1	$\{0,1\}$	1	$\{1,2\}$
2	$\{0,2\}$	$\{1,2\}$	2

$*$	a	b
a	a	$\{a,b\}$
b	$\{a,b\}$	b

and suppose that $f : H_1 \longrightarrow H_2$ is defined by $f(0) = f(1) = a$ and $f(2) = b$. Then, f is a good homomorphism of type 4.

On a hypergroup H, we are concerened with equivalence relations for which the family of equivalence classes forms a hypergroup under the hyperopertation induced by that on H. For an equivalence relation ρ on H, we may use x_ρ, $\overline{x}$ or $\rho(x)$ to denote the equivalence class of $x \in H$. Moreover, generally, if A is a non-empty subset of H, then $A_\rho = \cup\{x_\rho \mid x \in A\}$.

We let H/ρ (read H modulo ρ) denote the family $\{x_\rho \mid x \in H\}$ of classes of ρ. The hyperoperation on H induces a hyperoperation $\otimes$ on H/ρ defined by

$$x_\rho \otimes y_\rho = \{z_\rho \mid z \in x_\rho \circ y_\rho\},$$

where $x, y \in H$. The structure $(H/\rho, \otimes)$ is known as a *factor* or *quotient* *structure*. Note that in the definition of $\otimes$, the condition $z \in x_\rho \circ y_\rho$ maybe replaced by $z \in (x_\rho \circ y_\rho)_\rho$ or $z_\rho \subseteq (x_\rho \circ y_\rho)_\rho$. Obviously, $\cup(x_\rho \otimes y_\rho) = (x_\rho \circ y_\rho)_\rho$.

Proposition 2.4.10. *Let $(H, \circ)$ be a hypergroup. Then, $(H/\rho, \otimes)$ is a hypergroup if and only if for all $x, y, z \in H$,*

$$\left((x_\rho \circ y_\rho)_\rho \circ z_\rho\right)_\rho = \left(x_\rho \circ (y_\rho \circ z_\rho)_\rho\right)_\rho.$$

Proof. In H/ρ, we have

$$\begin{aligned}
(x_\rho \otimes y_\rho) \otimes z_\rho &= \{u_\rho \mid u \in x_\rho \circ y_\rho\} \otimes z_\rho \\
&= \{t_\rho \mid t \in u_\rho \circ z_\rho,\ u \in x_\rho \circ y_\rho\} \\
&= \{t_\rho \mid t \in (x_\rho \circ y_\rho)_\rho \circ z_\rho\}.
\end{aligned}$$

Similarly, $x_\rho \otimes (y_\rho \otimes z_\rho) = \{t_\rho \mid t \in x_\rho \circ (y_\rho \circ z_\rho)_\rho\}$. Therefore, $\otimes$ is associative. Reproducibility in $(H/\rho, \otimes)$ is a consequence of reproducibility in H. Suppose that $x_\rho,\ y_\rho \in H/\rho$. Let $u, v \in H$ such that $y \in x \circ u$, $v \circ x$. Then, obviousely, $y_\rho \in x_\rho \otimes u_\rho$, $v_\rho \otimes x_\rho$. Hence, the proposition holds. ∎

Definition 2.4.11. Let ρ be an equivalence relation on a hypergroup $(H, \circ)$. Then, given $x, y \in H$, ρ is said to be of

> *type 1*, if H/ρ is a hypergroup;
> *type 2*, if $x_\rho \circ y_\rho \subseteq (x \circ y)_\rho$;
> *type 3*, if $(x \circ y)_\rho \subseteq x_\rho \circ y_\rho$;
> *type 4*, if $x_\rho \circ y_\rho = (x \circ y)_\rho$.

Observe that being of type 2 is equivalent to $(x_\rho \circ y_\rho)_\rho = (x \circ y)_\rho$, and of type 3 to $(x_\rho \circ y_\rho)_\rho = x_\rho \circ y_\rho$. Note that for an equivalence of type 2, $x_\rho \otimes y_\rho = \{z_\rho \mid z \in x \circ y\}$.

Proposition 2.4.12. *Let ρ be an equivalence relation on a hypergroup $(H, \circ)$. Then, ρ is of*

> *(1) type 4 if and only if ρ is of type 2 and type 3;*
> *(2) type 1 if ρ is of type 2 or type 3.*

Proof. (1) It is straightforward.

(2) Let $x, y, z \in H$. Suppose that ρ is of type 2. Then, we obtain

$$\left((x_\rho \circ y_\rho)_\rho \circ z_\rho\right)_\rho = ((x \circ y) \circ z)_\rho \quad \text{and} \quad \left(x_\rho \circ (y_\rho \circ z_\rho)_\rho\right)_\rho = (x \circ (y \circ z))_\rho.$$

Suppose that ρ is of type 3. Then, we have

$$\left((x_\rho \circ y_\rho)_\rho \circ z_\rho\right)_\rho = (x_\rho \circ y_\rho) \circ z_\rho \quad \text{and} \quad \left(x_\rho \circ (y_\rho \circ z_\rho)_\rho\right)_\rho = x_\rho \circ (y_\rho \circ z_\rho).$$

Hence, in either case, Proposition 2.4.10 applies and yields that H/ρ is a hypergroup. Therefore, ρ is of type 1 and (2) holds. ∎

Homomorphisms between hypergroups and equivalence relations on hypergroups are closely related. The following results are fundamental.

Theorem 2.4.13. *Let $(H_1, \circ)$ and $(H_2, *)$ be two hypergroups and $f :$ $H_1 \longrightarrow H_2$ be an onto mapping. We denote also by f the equivalence relation by f on H_1 whose classes comprise the family $\{x_f \mid x \in H_1\}$. Then, for $n = 1, 2, 3$ or 4, f is an equivalence relation of type n on H_1 for which H_1/f is canonically isomorphic to H_2 if and only if f is a homomorphism of type n.*

Proof. Let $n = 1$. Suppose that f is a homomorphism of type 1. In order to show f is an equivalence relation of type 1 on H_1, Proposition 2.4.10 is employed. Let $x, y, z \in H_1$. By Corollary 2.4.8 (1),

$$
\begin{aligned}
((x_f \circ y_f)_f \circ z_f)_f &= f^{-1} \left(f((x_f \circ y_f)_f \circ z_f) \right) \\
&= f^{-1}(f(x_f \circ y_f) * f(z)) \\
&= f^{-1}(f(x) * f(y) * f(z)).
\end{aligned}
$$

Similarly, we obtain $(x_f \circ (y_f \circ z_f)_f)_f = f^{-1}(f(x) * f(y) * f(z))$. Thus, f is an equivalence relation of type 1. The canonical mapping θ of H_1/f onto H_2 is given for $x \in H_1$ by $\theta(x_f) = f(x)$. It is clearly well defined and one to one. Moreover, for $x, y \in H_1$,

$$
\theta(x_f \otimes y_f) = \theta(\{z_f \mid z \in x_f \circ y_f\}) = \{f(z) \mid z \in x_f \circ y_f\} = f(x_f \circ y_f)
$$

and

$$
\theta(x_f) * \theta(y_f) = f(x) * f(y).
$$

Therefore, Proposition 2.4.7 (1) yields that H_1/f is canonically isomorphic to H_2 if and only if f is a homomorphism of type 1. The theorem is then established for $n = 1$.

Now, let $n > 1$. If f is a homomorphism of type n, then Proposition 2.4.6 and the theorem for $n = 1$ imply that H_1/f is canonically isomorphic to H_2. On the other hand, if H_1/f is canonically isomorphic to H_2, then the above relations yields for $x, y \in H_1$ that $f^{-1}(f(x) * f(y)) = (x_f \circ y_f)_f$. Therefore, for $n = 2, 3$ or 4, f is an equivalence relation of type n if and only if f is a homomorphism of type n. ∎

Corollary 2.4.14. *Let $(H, \circ)$ be a hypergroup. Let θ be an equivalence relation on H. We denote also by θ the canonical mapping of H onto H/θ. Then, for $n = 1, 2, 3$ or 4, θ is an equivalence relation of type n on H if and only if H/θ is a hypergroup and θ is a homomorphism of type n of H onto H/θ.*

Proof. By Proposition 2.4.13, if θ is an equivalence relation of type 1, 2, 3 or 4, then H/θ is a hypergroup. Note that for $x \in H$, $\theta(x) = x_\theta$, an element of H/θ, and that $\theta^{-1}(\theta(x)) = x_\theta$, a subset of H. This compatibility allows the theorem to be applied and yields the corollary. ∎

Three equivalence relations on a hypergroup are of such importance that they will be referred to as the fundamental equivalences. They arise as one tries to discriminate between pairs of elements by means of the hyperoperation.

Definition 2.4.15. Let $(H, \circ)$ be a hypergroup. Let $x, y \in H$. Then, x and y are said to be *operationally equivalent* or *0-equivalent* if $x \circ a = y \circ a$ and $a \circ x = a \circ y$ for every $a \in H$. The elements x and y are said to be *inseparable* or *i-equivalent* if $x \in a \circ b$ when and only when $y \in a \circ b$ for every $a, b \in H$. Also, x and y are said to be *essentially indistinguishable* or *e-equivalent* if they are both operationally equivalent and inseparable.

Obviously, the relations *o-*, *i-* and *e*-equivalence, denoted respectively by o, i, e, are equivalence relations on a hypergroup H. For $x \in H$, the *o-*, *i-* and *e*-equivalence classes of x are hence denoted by x_o, x_i and x_e, respectively.

Proposition 2.4.16. *Let $(H, \circ)$ be a hypergroup and $x, y \in H$.*

(1) $x_o \circ y_o = x \circ y$; o-equivalence is of type 2.
(2) $(x \circ y)_i = x \circ y$; i-equivalence is of type 3.
(3) $x_e = x_o \cap x_i$; $x_e \circ y_e = (x \circ y)_e = x \circ y$; e-equivalence is of type 4.

Proof. It is an immediate consequence of definition. ∎

Corollary 2.4.17. *Given that H is a hypergroup, H/o, H/i and H/e are hypergroups. The canonical mappings of H onto H/o, H/i and H/e are homomorphisms of type $2, 3$ and 4, respectively.*

Definition 2.4.18. A hypergroup H in which $x_o = x$ for each $x \in H$, $x_i = x$ for each $x \in H$ or $x_e = x$ for each $x \in H$ is said respectively to be *o-reduced*, *i-reduced* or *e-reduced*. An *e*-reduced hypergroup is simply said to be *reduced*.

Example 2.4.19. Let $H = \{a, b, c, d\}$. Let the hyperoperation $\circ$ on H be given by the following table:

$\circ$	a	b	c	d
a	$\{a, b\}$	$\{a, b\}$	$\{c, d\}$	$\{c, d\}$
b	$\{a, b\}$	$\{a, b\}$	$\{c, d\}$	$\{c, d\}$
c	$\{c, d\}$	$\{c, d\}$	a	b
d	$\{c, d\}$	$\{c, d\}$	b	a

One easily checks that $(H, \circ)$ is a hypergroup. The o-equivalence classes are seen to be $\{a, b\}$, c and d, whereas the i-equivalence classes are a, b and $\{c, d\}$. Therefore, the e-equivalence classes are all singletons, that is to say, H is e-reduced.

Example 2.4.20. Let $H = \mathbb{Z} \times \mathbb{Z}^*$, where $\mathbb{Z}^*$ is the set of non-zero integers. Let ρ be the equivalence relation that puts equivalent "fraction" into classes, that is, for $(x, y) \in H$,

$$(x, y)_\rho = \{(u, v) \mid xv = yu\}.$$

The hyperoperation $\circ$ on H is given for $(w, x), (y, z) \in H$ by

$$(w, x) \circ (y, z) = (wz + xy, xz)_\rho.$$

Then, $(H, \circ)$ is a hypergroup in which

$$(x, y)_o = (x, y)_i = (x, y)_e = (x, y)_\rho,$$

for $(x, y) \in H$. Furthermore, $(H/\rho, \otimes) \cong (\mathbb{Q}, +)$.

Proposition 2.4.21. *Let $(H_1, \circ)$, $(H_2, *)$ and $(H_3, \bullet)$ be hypergroups. For $n = 1, 2, 3$ or 4, let f be a homomorphism of type n of H_1 onto H_2 and g be a homomorphism of type n of H_2 onto H_3. Then, gf is a homomorphism of type n of H_1 onto H_3.*

Proof. Suppose that $x, y \in H_1$. We first observe that for $z \in H_1$,

$$z_{gf} = f^{-1}g^{-1}gf(z) = f^{-1}(f(z)_g).$$

Let $n = 1$. By the above relation, we obtain

$$gf(x_{gf} \circ y_{gf}) = gf\left(f^{-1}(f(x)_g) \circ f^{-1}(f(y)_g)\right).$$

Since f is onto, Corollary 2.4.8 (1) applies and yields

$$gf\left(f^{-1}(f(x)_g) \circ f^{-1}(f(y)_g)\right) = g(f(x)_g * f(y)_g).$$

Then, Proposition 2.4.7 (1) gives

$$g\left(f(x)_g * f(y)_g\right) = gf(x) \bullet gf(y).$$

Hence, gf is a homomorphism of type 1 by the above relations and Proposition 2.4.7 (1).

Let $n = 2$. Similar to the previous case, but simpler.

Let $n = 3$. Since g is of type 3,

$$f^{-1}g^{-1}(gf(x) \bullet gf(y)) = f^{-1}(f(x)_g * f(y)_g).$$

Since f is onto, Proposition 2.4.5 (3) applies and gives

$$f^{-1}(f(x)_g * f(y)_g) = f^{-1}(f(x)_g) \circ f^{-1}(f(y)_g).$$

Now, we obtain

$$f^{-1}(f(x)_g) \circ f^{-1}(f(y)_g) = x_{gf} \circ y_{gf}.$$

So, by the above relations, we conclude that gf is a homomorphism of type 3.

Let $n = 4$. Then, gf is a homomorphism of type 4 by Proposition 2.4.6 (1). ∎

In the following definition, we introduce more types of homomorphisms of hypergroups that appeared in the literature under various names.

For a, b, we denote $a/b = \{x \mid a \in x \circ b\}$ and $b \backslash a = \{y \mid a \in b \circ y\}$.

Definition 2.4.22. Let $(H_1, \circ)$ and $(H_2, *)$ be two hypergroups and $f : H_1 \longrightarrow H_2$ be a mapping. We say that f is a homomorphism of

> *type 5*, if for all $x, y \in H_1$, f is a good homomorphism and furthermore
>
> $$(1)\ f(x/y) = f(x/f^{-1}(f(y))),$$
> $$(2)\ f(y \backslash x) = f(f^{-1}(f(y)) \backslash x);$$
>
> *type 6*, if for all $x, y \in H_1$, f is a good homomorphism and furthermore
>
> $$(3)\ f(x/f^{-1}(f(y))) = f(x)/f(y),$$
> $$(4)\ f(f^{-1}(f(y)) \backslash x) = f(y) \backslash f(x);$$
>
> *type 7*, if for all $x, y \in H_1$, f is a good homomorphism and furthermore
>
> $$(5)\ f(x/y) = f(x)/f(y),$$
> $$(6)\ f(y \backslash x) = f(y) \backslash f(x).$$

Theorem 2.4.23. *If $f : H_1 \longrightarrow H_2$ is a homomorphism of type 7, then f is a homomorphism of type 4.*

Proof. In general, if f is an inclusion homomorphism, then for every $x, y \in H_1$,

$$f^{-1}(f(x)) \circ f^{-1}(f(y)) \subseteq f^{-1}(f(x) * f(y)).$$

Now, let f be a homomorphism of type 7. Suppose that $z \in f^{-1}(f(x) * f(y))$. Then, $f(z) \in f(x) * f(y)$, which implies that $f(y) \in f(x) \backslash f(z)$ and so $f(y) \in f(x \backslash z)$. Thus, there exists $y' \in x \backslash z$ such that $f(y) = f(y')$, consequently $z \in x \circ y' \subseteq f^{-1}(f(x)) \circ f^{-1}(f(y))$. Therefore, $f^{-1}(f(x) * f(y)) \subseteq f^{-1}(f(x)) \circ f^{-1}(f(y))$. ∎

Theorem 2.4.24. *If $f : H_1 \longrightarrow H_2$ is an onto homomorphism of type 4, then f is a homomorphism of type 6.*

Proof. We know that an onto homomorphism of type 4 is a good homomorphism. Suppose that $u \in f(z)/f(x)$. Then, there exists y such that $f(y) = u$ and so $f(z) \in f(y) * f(x)$. Consequently, $z \in f^{-1}(f(y) * f(x)) = f^{-1}(f(y)) \circ f^{-1}(f(x))$, it follows that there exist a, b such that $z \in a \circ b$, where $a \in f^{-1}(f(y))$ and $b \in f^{-1}(f(x))$. Hence, $f(y) = f(a)$, $f(x) = f(b)$ and so $u = f(a) \in f(z/b) \subseteq f(z/f^{-1}(f(x)))$. Therefore, $f(z)/f(x) \subseteq f(z/f^{-1}(f(x)))$. Note that the inverse inclusion is always true. Similarly, we can prove $f(f^{-1}(f(y)) \backslash x) = f(y) \backslash f(x)$. ∎

Notice that, in general, a homomorphism of type 4 is not good.

Example 2.4.25. Let $\mathcal{H}(\mathbb{Z})$ be the hypergroup $(\mathbb{N}, \circ)$, where for all x, y,

$$x \circ y = \{x + y, \ |x - y|\}.$$

Let $f : \mathcal{H}(\mathbb{Z}) \longrightarrow \mathcal{H}(\mathbb{Z})$ be the function defined by

$$f(x) = \begin{cases} 0 \text{ if } x \in 2\mathbb{N} \\ 1 \text{ if } x \in 2\mathbb{N} + 1. \end{cases}$$

Then, f is a homomorphism of type 4 but is not a good homomorphism.

2.5 Regular and strongly regular relations

By using a certain type of equivalence relations, we can connect semihypergroups to semigroups and hypergroups to groups. These equivalence relations are called strong regular relations. More exactly, by a given (semi)hypergroup and by using a strong regular relation on it, we can construct a (semi)group structure on the quotient set. A natural question arises: Do they also exist regular relations? The answer is positive, regular relations provide us new (semi)hypergroup structures on the quotient sets.

Let us define these notions. First, we do some notations.

Let $(H, \circ)$ be a semihypergroup and R be an equivalence relation on H. If A and B are non-empty subsets of H, then

$$A \overline{R} B \text{ means that } \forall a \in A, \exists b \in B \text{ such that } aRb \text{ and}$$
$$\forall b' \in B, \exists a' \in A \text{ such that } a' R b';$$
$$A \overline{\overline{R}} B \text{ means that } \forall a \in A, \forall b \in B, \text{ we have } aRb.$$

Definition 2.5.1. The equivalence relation R is called

(1) *regular on the right* (*on the left*) if for all x of H, from aRb, it follows that $(a \circ x)\overline{R}(b \circ x)$ $((x \circ a)\overline{R}(x \circ b)$ respectively);
(2) *strongly regular on the right* (*on the left*) if for all x of H, from aRb, it follows that $(a \circ x)\overline{\overline{R}}(b \circ x)$ $((x \circ a)\overline{\overline{R}}(x \circ b)$ respectively);
(3) R is called *regular* (*strongly regular*) if it is regular (strongly regular) on the right and on the left.

Theorem 2.5.2. *Let* $(H, \circ)$ *be a semihypergroup and* R *be an equivalence relation on* H.

(1) *If* R *is regular, then* H/R *is a semihypergroup, with respect to the following hyperoperation:* $\overline{x} \otimes \overline{y} = \{\overline{z} \mid z \in x \circ y\}$;
(2) *If the above hyperoperation is well defined on* H/R, *then* R *is regular.*

Proof. (1) First, we check that the hyperoperation $\otimes$ is well defined on H/R. Consider $\overline{x} = \overline{x_1}$ and $\overline{y} = \overline{y_1}$. We check that $\overline{x} \otimes \overline{y} = \overline{x_1} \otimes \overline{y_1}$. We have xRx_1 and yRy_1. Since R is regular, it follows that $(x \circ y)\overline{R}(x_1 \circ y)$, $(x_1 \circ y)\overline{R}(x_1 \circ y_1)$ whence $(x \circ y)\overline{R}(x_1 \circ y_1)$. Hence, for all $z \in x \circ y$, there exists $z_1 \in x_1 \circ y_1$ such that zRz_1, which means that $\overline{z} = \overline{z_1}$. It follows that $\overline{x} \otimes \overline{y} \subseteq \overline{x_1} \otimes \overline{y_1}$ and similarly we obtain the converse inclusion. Now, we check the associativity of $\otimes$. Let $\overline{x}, \overline{y}, \overline{z}$ be arbitrary elements in H/R and $\overline{u} \in (\overline{x} \otimes \overline{y}) \otimes \overline{z}$. This means that there exists $\overline{v} \in \overline{x} \otimes \overline{y}$ such that $\overline{u} \in \overline{v} \otimes \overline{z}$. In other words, there exist $v_1 \in x \circ y$ and $u_1 \in v \circ z$, such that vRv_1 and uRu_1. Since R is regular, it follows that there exists $u_2 \in v_1 \circ z \subseteq x \circ (y \circ z)$ such that u_1Ru_2. From here, we obtain that there exists $u_3 \in y \circ z$ such that $u_2 \in x \circ u_3$. We have $\overline{u} = \overline{u_1} = \overline{u_2} \in \overline{x} \otimes \overline{u_3} \subseteq \overline{x} \otimes (\overline{y} \otimes \overline{z})$. It follows that $(\overline{x} \otimes \overline{y}) \otimes \overline{z} \subseteq \overline{x} \otimes (\overline{y} \otimes \overline{z})$. Similarly, we obtain the converse inclusion.

(2) Let aRb and x be an arbitrary element of H. If $u \in a \circ x$, then $\overline{u} \in \overline{a} \otimes \overline{x} = \overline{b} \otimes \overline{x} = \{\overline{v} \mid v \in b \circ x\}$. Hence, there exists $v \in b \circ x$ such that uRv, whence $(a \circ x)\overline{R}(b \circ x)$. Similarly we obtain that R is regular on the left. ∎

Corollary 2.5.3. *If* $(H, \circ)$ *is a hypergroup and* R *is an equivalence relation on* H, *then* R *is regular if and only if* $(H/R, \otimes)$ *is a hypergroup.*

Proof. If H is a hypergroup, then for all x of H we have $H \circ x = x \circ H = H$, whence we obtain $H/R \otimes \overline{x} = \overline{x} \otimes H/R = H/R$. According to the above theorem, it follows that $(H/R, \otimes)$ is a hypergroup. ∎

Notice that if R is regular on a (semi)hypergroup H, then the canonical projection $\pi : H \longrightarrow H/R$ is a good epimorphism. Indeed, for all x, y of H and $\overline{z} \in \pi(x \circ y)$, there exists $z' \in x \circ y$ such that $\overline{z} = \overline{z'}$. We have $\overline{z} = \overline{z'} \in \overline{x} \otimes \overline{y} = \pi(x) \otimes \pi(y)$. Conversely, if $\overline{z} \in \pi(x) \otimes \pi(y) = \overline{x} \otimes \overline{y}$, then there exists $z_1 \in x \circ y$ such that $\overline{z} = \overline{z_1} \in \pi(x \circ y)$.

Theorem 2.5.4. *If $(H, \circ)$ and $(K, *)$ are semihypergroups and $f : H \longrightarrow K$ is a good homomorphism, then the equivalence ρ^f associated with f, that is $x\rho^f y \Leftrightarrow f(x) = f(y)$, is regular and $\varphi : f(H) \longrightarrow H/\rho^f$, defined by $\varphi(f(x)) = \overline{x}$, is an isomorphism.*

Proof. Let $h_1 \rho^f h_2$ and a be an arbitrary element of H. If $u \in h_1 \circ a$, then
$$f(u) \in f(h_1 \circ a) = f(h_1) * f(a) = f(h_2) * f(a) = f(h_2 \circ a).$$

Then, there exists $v \in h_2 \circ a$ such that $f(u) = f(v)$, which means that $u\rho^f v$. Hence, ρ^f is regular on the right. Similarly, it can be shown that ρ^f is regular on the left. On the other hand, for all $f(x), f(y)$ of $f(H)$, we have

$$\varphi(f(x) * f(y)) = \varphi(f(x \circ y)) = \{\overline{z} \mid z \in x \circ y\} = \overline{x} \otimes \overline{y} = \varphi(f(x)) \otimes \varphi(f(y)).$$

Moreover, if $\varphi(f(x)) = \varphi(f(y))$, then $x\rho^f y$, so φ is injective and clearly, it is also surjective. Finally, for all $\overline{x}, \overline{y}$ of H/ρ^f we have

$$\begin{aligned} \varphi^{-1}(\overline{x} \otimes \overline{y}) &= \varphi^{-1}(\{\overline{z} \mid z \in x \circ y\}) = \{f(z) \mid z \in x \circ y\} \\ &= f(x \circ y) = f(x) * f(y) = \varphi^{-1}(\overline{x}) * \varphi^{-1}(\overline{y}). \end{aligned}$$

Therefore, φ is an isomorphism. ■

Theorem 2.5.5. *Let $(H, \circ)$ be a semihypergroup and R be an equivalence relation on H.*

(1) *If R is strongly regular, then H/R is a semigroup, with respect to the following operation: $\overline{x} \otimes \overline{y} = \overline{z}$, for all $z \in x \circ y$;*

(2) *If the above operation is well defined on H/R, then R is strongly regular.*

Proof. (1) For all x, y of H, we have $(x \circ y)\overline{\overline{R}}(x \circ y)$. Hence, $\overline{x} \otimes \overline{y} = \{\overline{z} \mid z \in x \circ y\} = \{\overline{z}\}$, which means that $\overline{x} \otimes \overline{y}$ has exactly one element. Therefore, $(H/R, \otimes)$ is a semigroup.

(2) If aRb and x is an arbitrary element of H, we check that $(a \circ x)\overline{\overline{R}}(b \circ x)$. Indeed, for all $u \in a \circ x$ and all $v \in b \circ x$ we have $\overline{u} = \overline{a} \otimes \overline{x} = \overline{b} \otimes \overline{x} = \overline{v}$, which means that uRv. Hence, R is strongly regular on the right and similarly, it can be shown that it is strongly regular on the left. ■

Corollary 2.5.6. *If $(H, \circ)$ is a hypergroup and R is an equivalence relation on H, then R is strongly regular if and only if $(H/R, \otimes)$ is a group.*
Proof. It is obvious. ∎

Theorem 2.5.7. *If $(H, \circ)$ is a semihypergroup, $(S, *)$ is a semigroup and $f : H \longrightarrow S$ is a homomorphism, then the equivalence ρ^f associated with f is strongly regular.*
Proof. Let $a \rho^f b$, $x \in H$ and $u \in a \circ x$. It follows that

$$f(u) = f(a) * f(x) = f(b) * f(x) = f(b \circ x).$$

Hence, for all $v \in b \circ x$, we have $f(u) = f(v)$, which means that $u \rho^f v$. Hence, ρ^f is strongly regular on the right and similarly, it is strongly regular on the left. ∎

The fundamental relation has an important role in the study of semi-hypergroups and especially of hypergroups.

Definition 2.5.8. For all $n > 1$, we define the relation β_n on a semihyper-group H, as follows:

$$a \, \beta_n \, b \Leftrightarrow \exists (x_1, \ldots, x_n) \in H^n : \{a, b\} \subseteq \prod_{i=1}^{n} x_i,$$

and $\beta = \bigcup_{n \geq 1} \beta_n$, where $\beta_1 = \{(x, x) \mid x \in H\}$ is the diagonal relation on H. Clearly, the relation β is reflexive and symmetric. Denote by β^* the transitive closure of β.

Theorem 2.5.9. β^* *is the smallest strongly regular relation on H.*

Proof. We show that:

(1) β^* is a strongly regular relation on H;
(2) If R is a strongly regular relation on H, then $\beta^* \subseteq R$.

(1) Let $a \, \beta^* b$ and x be an arbitrary element of H. It follows that there exist $x_0 = a, x_1, \ldots, x_n = b$ such that for all $i \in \{0, 1, \ldots, n-1\}$ we have $x_i \, \beta \, x_{i+1}$. Let $u_1 \in a \circ x$ and $u_2 \in b \circ x$. We check that $u_1 \, \beta^* \, u_2$. From $x_i \, \beta \, x_{i+1}$ it follows that there exists a hyperproduct P_i, such that $\{x_i, x_{i+1}\} \subseteq P_i$ and so $x_i \circ x \subseteq P_i \circ x$ and $x_{i+1} \circ x \subseteq P_i \circ x$, which means that $x_i \circ x \overline{\overline{\beta}} x_{i+1} \circ x$. Hence, for all $i \in \{0, 1, \ldots, n-1\}$ and for all $s_i \in x_i \circ x$ we have $s_i \, \beta \, s_{i+1}$. If we consider $s_0 = u_1$ and $s_n = u_2$, then we obtain $u_1 \, \beta^* \, u_2$. Then, β^* is strongly regular on the right and similarly, it is strongly regular on the left.

(2) We have $\beta_1 = \{(x,x) \mid x \in H\} \subseteq R$, since R is reflexive. Suppose that $\beta_{n-1} \subseteq R$ and show that $\beta_n \subseteq R$. If $a\beta_n b$, then there exist $x_1, \ldots, x_n$ in H, such that $\{a,b\} \subseteq \prod\limits_{i=1}^{n} x_i$. Hence, there exists u, v in $\prod\limits_{i=1}^{n-1} x_i$, such that $a \in u \circ x_n$ and $b \in v \circ x_n$. We have $u\beta_{n-1}v$ and according to the hypothesis, we obtain uRv. Since R is strongly regular, it follows that aRb. Hence, $\beta_n \subseteq R$. By induction, it follows that $\beta \subseteq R$, whence $\beta^* \subseteq R$. $\blacksquare$

Hence, the relation β^* is the smallest equivalence relation on H, such that the quotient H/β^* is a group.

Definition 2.5.10. β^* is called the *fundamental equivalence relation* on H and H/β^* is called the *fundamental group*.

If H is a hypergroup, then $\beta = \beta^*$ [74]. Consider the canonical projection $\varphi_H : H \longrightarrow H/\beta^*$. The *heart* of H is the set $\omega_H = \{x \in H \mid \varphi_H(x) = 1\}$, where 1 is the identity of the group H/β^*. This relation was introduced by Koskas [94] and studied mainly by Corsini, Davvaz, Freni, Leoreanu, Vougiouklis and many others.

Freni in [73] introduced the relation γ as a generalization of the relation β.

Definition 2.5.11. Let H be a semihypergroup. Then, we set

$$\gamma_1 = \{(x,x) \mid x \in H\}$$

and for every integer $n > 1$, γ_n is the relation defined as follows:

$$x \, \gamma_n \, y \Longleftrightarrow \exists (z_1, \ldots, z_n) \in H^n, \ \exists \sigma \in \mathbb{S}_n \ : \ x \in \prod\limits_{i=1}^{n} z_i, \ y \in \prod\limits_{i=1}^{n} z_{\sigma(i)}.$$

Obviously, for $n \geq 1$, the relations γ_n are symmetric, and the relation $\gamma = \bigcup\limits_{n\geq 1} \gamma_n$ is reflexive and symmetric.

Let γ^* be the transitive closure of γ. If H is a hypergroup, then $\gamma = \gamma^*$ [73].

Theorem 2.5.12. *The relation γ^* is a strongly regular relation.*

Proof. Clearly, γ^* is an equivalence relation. In order to prove that it is strongly regular, we have to show first that

$$x\gamma y \Longrightarrow (x \circ a) \, \overline{\overline{\gamma}} \, (y \circ a) \text{ and } (a \circ x) \, \overline{\overline{\gamma}} \, (a \circ y),$$

for every $a \in H$. If $x\gamma y$, then there is $n \in \mathbb{N}$ such that $x\gamma_n y$. Hence, there exist $(z_1, \ldots, z_n) \in H^n$ and $\sigma \in \mathbb{S}_n$ such that $x \in \prod\limits_{i=1}^{n} z_i$ and $y \in \prod\limits_{i=1}^{n} z_{\sigma(i)}$.

For every $a \in H$, set $a = z_{n+1}$ and let τ be a permutation of $\mathbb{S}_{n+1}$ such that

$$\tau(i) = \sigma(i), \forall i \in \{1, 2, \ldots, n\};$$
$$\tau(n+1) = n+1.$$

For all $v \in x \circ a$ and for all $w \in y \circ a$, we have $v \in x \circ a \subseteq \prod_{i=1}^{n} z_i \circ a = \prod_{i=1}^{n+1} z_i$

and $w \in y \circ a \subseteq \prod_{i=1}^{n} z_{\sigma(i)} \circ a = \prod_{i=1}^{n} z_{\sigma(i)} \circ z_{n+1} = \prod_{i=1}^{n+1} z_{\tau(i)}$. So, $v \, \gamma_{n+1} \, w$ and

hence $v \, \gamma \, w$. Thus, $(x \circ a) \, \overline{\overline{\gamma}} \, (y \circ a)$. In the same way, we can show that $(a \circ x) \, \overline{\overline{\gamma}} \, (a \circ y)$.

Moreover, if $x \, \gamma^* \, y$, then there exist $m \in \mathbb{N}$ and

$$(w_0 = x, w_1, \ldots, w_{m-1}, w_m = y) \in H^{m+1}$$

such that $x = w_0 \, \gamma \, w_1 \gamma \ldots \gamma \, w_{m-1} \, \gamma \, w_m = y$. Now, we obtain

$$x \circ a = w_0 \circ a \, \overline{\overline{\gamma}} \, w_1 \circ a \, \overline{\overline{\gamma}} \, w_2 \circ a \, \overline{\overline{\gamma}} \, \ldots \, \overline{\overline{\gamma}} \, w_{m-1} \circ a \, \overline{\overline{\gamma}} \, w_m \circ a = y \circ a.$$

Finally, for all $v \in x \circ a = w_0 \circ a$ and for all $w \in w_m \circ a = y \circ a$, taking $z_1 \in w_1 \circ a$, $z_2 \in w_2 \circ a, \ldots, z_{m-1} \in w_{m-1} \circ a$, we have $v \, \gamma \, z_1 \, \gamma \, z_2 \, \gamma \ldots \gamma \, z_{m-1} \, \gamma \, w$, and so $v \, \gamma^* \, w$. Therefore, $x \circ a \overline{\overline{\gamma^*}} \, y \circ a$. Similarly, we obtain $a \circ x \overline{\overline{\gamma^*}} \, a \circ y$. Hence, γ^* is strongly regular. ∎

Corollary 2.5.13. *The quotient H/γ^* is a commutative semigroup. Furthermore, if H is a hypergroup, then H/γ^* is a commutative group.*

Proof. Since γ^* is a strongly regular relation, the quotient H/γ^* is a semigroup under the following operation:

$$\gamma^*(x_1) \otimes \gamma^*(x_2) = \gamma^*(z), \text{ for all } z \in x_1 \circ x_2.$$

Moreover, if H is a hypergroup, then H/γ^* is a group. Finally, if σ is the cycle of $\mathbb{S}_2$ such that $\sigma(1) = 2$, for all $z \in x_1 \circ x_2$ and $w \in x_{\sigma(1)} \circ x_{\sigma(2)}$, we have $z\gamma_2 w$, so $z\gamma^* w$ and $\gamma^*(x_1) \otimes \gamma^*(x_2) = \gamma^*(z) = \gamma^*(x_2) \otimes \gamma^*(x_1)$. ∎

Theorem 2.5.14. *The relation γ^* is the smallest strongly regular relation on a semihypergroup H such that the quotient H/γ^* is commutative semigroup.*

Proof. Suppose that ρ is a strongly regular relation such that H/ρ is a commutative semigroup and $\varphi : H \longrightarrow H/\rho$ is the canonical projection. Then, φ is a good homomorphism. Moreover, if $x \, \gamma_n \, y$, then there exist $(z_1, \ldots, z_n) \in H^n$ and $\sigma \in \mathbb{S}_n$ such that $x \in \prod_{i=1}^{n} z_i$ and $y \in \prod_{i=1}^{n} z_{\sigma(i)}$, whence

$\varphi(x) = \varphi(z_1) \otimes \ldots \otimes \varphi(z_n)$ and $\varphi(y) = \varphi(z_{\sigma(1)}) \otimes \ldots \otimes \varphi(z_{\sigma(n)})$. By the commutativity of H/ρ, it follows that $\varphi(x) = \varphi(y)$ and $x \rho y$. Thus, $x \gamma_n y$ implies $x \rho y$, and obviously, $x \gamma y$ implies that $x \rho y$.

Finally, if $x \gamma^* y$, then there exist $m \in \mathbb{N}$ and

$$(w_0 = x, w_1, \ldots, w_{m-1}, w_m = y) \in H^{m+1}$$

such that $x = w_0 \gamma w_1 \gamma \ldots \gamma w_{m-1} \gamma w_m = y$. Therefore,

$$x = w_0 \rho w_1 \rho \ldots \rho w_{m-1} \rho w_m = y,$$

and transitivity of ρ implies that $x \rho y$. Therefore, $\gamma^* \subseteq \rho$. ∎

Complete parts were introduced and studied for the first time by M. Koskas [94]. Later, this topic was analyzed by P. Corsini [29], Y. Sureau [138] and B. Davvaz et al [54; 55] mostly in the general theory of hypergroups. M. De Salvo studied complete parts from a combinatorial point of view. A generalization of them, called n-complete parts, was introduced by R. Migliorato. Other authors gave a contribution to the study of complete parts and of the heart of a hypergroup. Among them, V. Leoreanu analyzed the structure of the heart of a hypergroup in her Ph.D. Thesis.

We present now the definitions.

Definition 2.5.15. Let $(H, \circ)$ be a semihypergroup and A be a non-empty subset of H. We say that A is a *complete part* of H if for any non-zero natural number n and for all $a_1, \ldots, a_n$ of H, the following implication holds:

$$A \cap \prod_{i=1}^{n} a_i \neq \emptyset \implies \prod_{i=1}^{n} a_i \subseteq A.$$

Theorem 2.5.16. *If $(H, \circ)$ is a semihypergroup and R is a strongly regular relation on H, then for all z of H, the equivalence class of z is a complete part of H.*

Proof. Let $a_1, \ldots, a_n$ be elements of H, such that

$$\bar{z} \cap \prod_{i=1}^{n} a_i \neq \emptyset.$$

Then, there exists $y \in \prod_{i=1}^{n} a_i$, such that $y R z$. The homomorphism $\pi : H \longrightarrow H/R$ is good and H/R is a semigroup. It follows that $\pi(z) = \pi(y) = \pi\left(\prod_{i=1}^{n} a_i\right) = \prod_{i=1}^{n} \pi(a_i)$. This means that $\prod_{i=1}^{n} a_i \subseteq \bar{z}$. ∎

Now, we want to determine some necessary and sufficient conditions so that the relation γ is transitive.

Definition 2.5.17. Let M be a non-empty subset of H. We say that M is a γ-*part* of H if for any non-zero natural number n, for all $(z_1, \ldots, z_n) \in H^n$ and for all $\sigma \in \mathbb{S}_n$, we have

$$M \cap \prod_{i=1}^{n} z_i \neq \emptyset \implies \prod_{i=1}^{n} z_{\sigma(i)} \subseteq M.$$

Lemma 2.5.18. *Let M be a non-empty subsets of H. Then, the following conditions are equivalent:*

(1) M is a γ-part of H;
(2) $x \in M$, $x\gamma y \implies y \in M$;
(3) $x \in M$, $x\gamma^* y \implies y \in M$.

Proof. $(1 \implies 2)$ If $(x, y) \in H^2$ is a pair such that $x \in M$ and $x \gamma y$, then there exist $n \in \mathbb{N}$, $(z_1, \ldots, z_n) \in H^n$ and $\sigma \in \mathbb{S}_n$ such that $x \in M \cap \prod_{i=1}^{n} z_i$ and $y \in \prod_{i=1}^{n} z_{\sigma(i)}$. Since M is a γ-part of H, we have $\prod_{i=1}^{n} z_{\sigma(i)} \subseteq M$ and $y \in M$.

$(2 \implies 3)$ Assume that $(x, y) \in H^2$ such that $x \in M$ and $x \gamma^* y$. Obviously, there exist $m \in \mathbb{N}$ and $(w_0 = x, w_1, \ldots, w_{m-1}, w_m = y) \in H^{m+1}$ such that $x = w_0 \gamma w_1 \gamma \ldots \gamma w_{m-1} \gamma w_m = y$. Since $x \in M$, applying (2) m times, we obtain $y \in M$.

$(3 \implies 1)$ Suppose that $M \cap \prod_{i=1}^{n} x_i \neq \emptyset$ and $x \in M \cap \prod_{i=1}^{n} x_i$. For every $\sigma \in \mathbb{S}_n$ and for every $y \in \prod_{i=1}^{n} x_{\sigma(i)}$, we have $x \gamma y$. Thus, $x \in M$ and $x \gamma^* y$. Finally, by (3), we obtain $y \in M$, whence $\prod_{i=1}^{n} x_{\sigma(i)} \subseteq M$. ∎

Before proving the next theorem, we introduce the following notations: Let H be a semihypergroup. For all $x \in H$, we set

- $T_n(x) = \left\{ (x_1, \ldots, x_n) \in H^n \mid x \in \prod_{i=1}^{n} x_i \right\}$;
- $P_n(x) = \bigcup \left\{ \prod_{i=1}^{n} x_{\sigma(i)} \mid \sigma \in \mathbb{S}_n, \ (x_1, \ldots, x_n) \in T_n(x) \right\}$;
- $P_\sigma(x) = \bigcup_{n \geq 1} P_n(x).$

From the preceding notations and definitions, it follows at once the following:

Lemma 2.5.19. *For every* $x \in H$, $P_\sigma(x) = \{y \in H \mid x \gamma y\}$.

Proof. For all $x, y \in H$, we have

$$x \gamma y \iff \exists n \in \mathbb{N}, \exists (x_1, \ldots, x_n) \in H^n, \exists \sigma \in \mathbb{S}_n : x \in \prod_{i=1}^{n} x_i, \ y \in \prod_{i=1}^{n} x_{\sigma(i)}$$

$$\iff \exists n \in \mathbb{N} : y \in P_n(x)$$

$$\iff y \in P_\sigma(x). \ \blacksquare$$

Theorem 2.5.20. *Let H be a semihypergroup. Then, the following conditions are equivalent:*

(1) γ is transitive;
(2) $\gamma^*(x) = P_\sigma(x)$, for all $x \in H$;
(3) $P_\sigma(x)$ is a γ-part of H, for all $x \in H$.

Proof. $(1 \Longrightarrow 2)$ By Lemma 2.5.19, for all $x, y \in H$, we have

$$y \in \gamma^*(x) \iff x \gamma^* y \iff x \gamma y \iff y \in P_\sigma(x).$$

$(2 \Longrightarrow 3)$ By Lemma 2.5.18, if M is a non-empty subset of H, then M is a γ-part of H if and only if it is union of equivalence classes modulo γ^*. In particular, every equivalence class modulo γ^* is a γ-part of H.

$(3 \Longrightarrow 1)$ If $x \ \gamma \ y$ and $y \ \gamma \ z$, then there exist $m, n \in \mathbb{N}$, $(x_1, \ldots, x_n) \in T_n(x)$, $(y_1, \ldots, y_m) \in T_m(y)$, $\sigma \in \mathbb{S}_n$ and $\tau \in \mathbb{S}_m$ such that $y \in \prod_{i=1}^{n} x_{\sigma(i)}$ and $z \in \prod_{i=1}^{m} y_{\tau(i)}$. Since $P_\sigma(x)$ is a γ-part of H, we have

$$x \in \prod_{i=1}^{n} x_i \cap P_\sigma(x) \Longrightarrow \prod_{i=1}^{n} x_{\sigma(i)} \subseteq P_\sigma(x) \Longrightarrow y \in \prod_{i=1}^{m} y_i \cap P_\sigma(x)$$

$$\Longrightarrow \prod_{i=1}^{m} y_{\tau(i)} \subseteq P_\sigma(x) \Longrightarrow z \in P_\sigma(x)$$

$$\Longrightarrow \exists k \in \mathbb{N} : z \in P_k(x) \Longrightarrow z \ \gamma \ x.$$

Therefore, γ is transitive. $\blacksquare$

2.6   Complete hypergroups

In this section, we study the concept of complete hypergroups.

Definition 2.6.1. Let A be a non-empty subset of H. The intersection of the parts of H which are complete and contain A is called the *complete closure* of A in H; it will be denoted by $C(A)$. A semihypergroup H is *complete*, if it satisfies one of the following conditions:

(1) $\forall (x, y) \in H^2$, $\forall a \in x \circ y$, $C(a) = x \circ y$.

(2) $\forall (x, y) \in H^2$, $C(x \circ y) = x \circ y$.

(3) $\forall (m, n) \in \mathbb{N}^2$, $2 \leq m, n$, $\forall (x_1, \ldots, x_n) \in H^n$, $\forall (y_1, \ldots, y_m) \in H^m$,

$$\prod_{i=1}^{n} x_i \cap \prod_{j=1}^{m} y_j \neq \emptyset \implies \prod_{i=1}^{n} x_i = \prod_{j=1}^{m} y_j.$$

A hypergroup is *complete* if it is a complete semihypergroup. If $(H, \circ)$ is a complete semihypergroup, then either there exist $a, b \in H$ such that $\beta^*(x) = a \circ b$ or $\beta^*(x) = \{x\}$.

An element $e \in H$ is called an *identity* if

$$a \in e \circ a \cap a \circ e \text{ for all } a \in H.$$

An element x' is called an *inverse* of x if an identity e exists such that

$$e \in x \circ x' \cap x' \circ x.$$

Definition 2.6.2. A *regular hypergroup* H is a hypergroup which it has at least one identity and every element has at least one inverse. A regular hypergroup H is said to be *reversible*, if it satisfies the following conditions:

$$\forall (a, b, x) \in H^3 : \quad a \in b \circ x \implies \exists x' \in i(x) : \quad b \in a \circ x',$$
$$a \in x \circ b \implies \exists x'' \in i(x) : \quad b \in x'' \circ a.$$

If H is regular, for every $x \in H$, we denote $i(x)$ the set of the inverses of x.

Theorem 2.6.3. *If $(H, \circ)$ is a complete hypergroup, then*

(1) $\omega_H = \{e \in H : \forall x \in H, x \in x \circ e \cap e \circ x\}$, *which means that ω_H is the set of two-sided identities of H.*

(2) *H is regular (i.e. H has at least one identity and any element has an inverse) and reversible.*

Proof. (1) If $u \in \omega_H$, then for all $a \in H$, we have $a \in C(a) = a \circ \omega_H = a \circ u$. Similarly we have $a \in u \circ a$, which means that u is a two-sided identity of H. Conversely, any two-sided identity u of H is an element of ω_H, since $\varphi(u) = 1$.

(2) Let a, a', a'' be elements of H and e be a two-sided identity, such that $e \in a' \circ a \cap a \circ a''$. Then, $a' \circ a = \omega_H = a \circ a''$ and $a \circ a' \subseteq a \circ a' \circ a \circ a'' \subseteq a \circ \omega_H \circ a'' = \omega_H \circ a \circ a'' = \omega_H$. Hence, $a \circ a' = \omega_H$, so a' is an inverse of a.

Moreover, if $a \in b \circ c$, then $\omega_H = a' \circ a \subseteq a' \circ b \circ c$, so for any inverse c' of c, we have

$$c' \in \omega_H \circ c' \subseteq a' \circ b \circ c \circ c' = a' \circ b \circ \omega_H = a' \circ b.$$

Similarly, from here we obtain $b' \in c \circ a'$, and so $b' \circ a \subseteq c \circ a' \circ a = C(c)$, whence $c \in C(c) = b' \circ a$. In a similar way, we obtain $b \in a \circ c'$. $\blacksquare$

Definition 2.6.4. A hypergroup $(H, \circ)$ is called *flat* if for all subhypergroup K of H, we have $\omega_K = \omega_H \cap K$.

Theorem 2.6.5. *Any complete hypergroup is flat.*

Proof. Let H be a complete hypergroup and let K be a subhypergroup H. We have $\omega_H \cap K = \{e \in K : \forall a \in H, \ x \in e \circ x \cap x \circ e\} \subseteq \omega_K$.

Moreover, $y \in C_K(x) \Rightarrow y\beta_K x \Rightarrow y\beta_H x \Rightarrow y \in C_H(x)$, which means that $C_K(x) \subseteq C_H(x)$. Clearly, $\omega_H \cap K \neq \emptyset$. If $x \in \omega_H \cap K \subseteq \omega_K$, then $C_K(x) = \omega_K, C_H(x) = \omega_H$. Hence, $\omega_K \subseteq \omega_H$ whence $\omega_K \subseteq \omega_H \cap K$. Hence, $\omega_K = \omega_H \cap K$. $\blacksquare$

Corollary 2.6.6. *If K is a subhypergroup of a complete hypergroup $(H, \circ)$, then $\omega_K = \omega_H$.*

Proof. Set $x \in \omega_H \cap K$. We have $\omega_H = C(x \circ x) = x \circ x \subseteq \omega_H \cap K$, whence $\omega_H \subseteq \omega_H \cap K$, then we apply the above theorem. Hence, $\omega_K = \omega_H$. $\blacksquare$

Theorem 2.6.7. *Let H, H' be complete hypergroups and $f : H \to H'$ be a good homomorphism. Then, we have $f(\omega_H) = \omega'_H$.*

Proof. Let $x \in \omega_H$. Then, $x \circ x = \omega_H$, whence $f(x) \circ f(x) = f(\omega_H)$. On the other hand, $f(x)$ is an identity of H', since x is an identity of H, which means that $f(x) \in \omega_{H'}$. Hence, $\omega'_H = f(x) \circ f(x) = f(\omega_H)$. $\blacksquare$

Now, we let $n \geq 2$.

Definition 2.6.8. A hypergroup H is said to be an *n-complete hypergroup* if for all $(z_1, \ldots, z_n) \in H^n$, $\prod_{i=1}^{n} z_i = \beta \left(\prod_{i=1}^{n} z_i \right)$.

If H is n-complete, then $\beta = \beta_n$.

A hypergroup H is said to be n^*-complete if there exists $n \in \mathbb{N}$ such that $\beta_n^* = \beta$ and $\beta_{n-1}^* \neq \beta_n^*$.

Definition 2.6.9. A hypergroup H is said to be γ_n-*complete* if for all $(z_1, z_2, \ldots, z_n) \in H^n$ and for all $\sigma \in \mathbb{S}_n$: $\gamma \left(\prod_{i=1}^{n} z_i \right) = \prod_{i=1}^{n} z_{\sigma(i)}$.

Corollary 2.6.10. *If H is a commutative hypergroup, then H is a γ_n-complete hypergroup if and only if H is an n-complete hypergroup.*

We begin with some properties which are valid in every hypergroup. They concern the relation γ_n and will be prove only for $n > 1$. The case

$n = 1$ is always trivially true. We suppose that $H = (H, \circ)$ is a hypergroup.

Proposition 2.6.11. *For any positive integer n, we have*

(1) $\gamma_n \subseteq \gamma_{n+1}$;

(2) $\gamma_n^* \subseteq \gamma_{n+1}^*$.

Proof. (1) If $x \; \gamma_n \; y$, then there exist $(z_1, z_2, \ldots, z_n) \in H^n$ and $\sigma \in \mathbb{S}_n$ such that $x \in \prod_{i=1}^{n} z_i$, $\quad y \in \prod_{i=1}^{n} z_{\sigma(i)}$. Since H is a hypergroup, so there exists $(t_1, t_2) \in H^2$ such that $z_n \in t_1 \circ t_2$.

Now let $z_i' = z_i$ for $1 \leq i \leq n-1$ and $z_n' = t_1, z_{n+1}' = t_2$, so $x \in z_1' \circ z_2' \ldots \circ z_n' \circ z_{n+1}'$. Now, let $\sigma(j) = n$ and we define $\sigma' \in \mathbb{S}_{n+1}$ as follows:

$$\sigma'(i) = \begin{cases} \sigma(i) & \text{if } 1 \leq i \leq j \\ n+1 & \text{if } i = j+1 \\ \sigma(i-1) & \text{if } j+2 \leq i. \end{cases}$$

Then, $\sigma' \in \mathbb{S}_{n+1}$ and $y \in \prod_{i=1}^{n+1} z_{\sigma'(i)}'$. Therefore, $x \; \gamma_{n+1} \; y$.

(2) It follows from (1). ∎

Proposition 2.6.12. *A hypergroup H is γ_n-complete if and only if for all $\sigma \in \mathbb{S}_n$ and for all $x \in \prod_{i=1}^{n} z_i$, we have $\gamma(x) = \prod_{i=1}^{n} z_{\sigma(i)}$.*

Proof. Suppose that H is γ_n-complete, for $\sigma \in \mathbb{S}_n$. If $x \in \prod_{i=1}^{n} z_i$, then we have

$$\gamma(x) \subseteq \bigcup_{x \in \prod_{i=1}^{n} z_i} \gamma(x) = \prod_{i=1}^{n} z_{\sigma(i)} \implies \gamma(x) \subseteq \prod_{i=1}^{n} z_{\sigma(i)}.$$

Now, if $y \in \prod_{i=1}^{n} z_{\sigma(i)}$, then

$$x \; \gamma_n \; y \implies x \; \gamma \; y \implies y \in \gamma(x) \implies \prod_{i=1}^{n} z_{\sigma(i)} \subseteq \gamma(x).$$

Conversely, for every $\sigma \in \mathbb{S}_n$, $(z_1, z_2, \ldots, z_n) \in H^n$ and $x \in \prod_{i=1}^{n} z_i$ we have

$$\gamma(x) = \prod_{i=1}^{n} z_{\sigma(i)} \implies \gamma(\prod_{i=1}^{n} z_i) = \bigcup_{x \in \prod_{i=1}^{n} z_i} \gamma(x) = \prod_{i=1}^{n} z_{\sigma(i)}. ∎$$

Proposition 2.6.13. *If H is a γ_n-complete hypergroup, then $\gamma^* = \gamma_n$.*

Proof. We know that in every hypergroup $\gamma^* = \gamma$, so it is suffices to prove that: $\gamma \subseteq \gamma_n$. Suppose that $x \gamma y$. Then, there exists $m \in \mathbb{N}$ such that $x \gamma_m y$. If $m \leq n$, then $\gamma_m \subseteq \gamma_n$. If $m > n$, then there exist $(z_1, z_2, \ldots, z_m) \in \dot{H}^m$ and $\sigma \in \mathbb{S}_m$ such that $x \in \prod_{i=1}^{m} z_i$, $y \in \prod_{i=1}^{m} z_{\sigma(i)}$. There exists $t_1 \in H$ such that $t_1 \in z_n \circ z_{n+1} \circ \ldots z_m$, $x \in \prod_{i=1}^{n-1} z_i \circ t_1$. Let $u_i = z_i$ for $1 \leq i \leq n-1$ and $u_n = t_1$. Thus, $x \in \prod_{i=1}^{n} u_i$. We have $y = \gamma(x) = \prod_{i=1}^{n} u_{\sigma(i)}$ which implies that $x \gamma_n y$. ∎

Example 2.6.14. If H is a semihypergroup, then Proposition 2.6.13 maybe not correct: let H be the following semihypergroup:

H	1	2	3	4
1	$1,2,3$	$1,2$	$1,3$	$1,3$
2	$1,2$	$1,2,3$	$2,3$	$1,3$
3	$1,3$	$2,3$	$1,2,3$	$1,2,3$
4	$1,3$	$1,3$	$1,2,3$	$1,2,3$

Then, H is γ_2-complete, $4\gamma^*4$ but not $4\gamma_2 4$.

Lemma 2.6.15. *For all $(a, b, x) \in H^3$,*
$$a \gamma_n b \Longrightarrow [(a \circ x) \overline{\overline{\gamma_{n+1}}} (b \circ x) \text{ and } (x \circ a) \overline{\overline{\gamma_{n+1}}} (x \circ b)].$$

Proof. If $a \gamma_n b$, then there exist $(z_1, z_2, \ldots, z_n) \in H^n$ and $\sigma \in \mathbb{S}_n$ such that $a \in \prod_{i=1}^{n} z_i$, $b \in \prod_{i=1}^{n} z_{\sigma(i)}$. But, we have $a \circ x \subseteq (\prod_{i=1}^{n} z_i) \circ x$ and $b \circ x \subseteq (\prod_{i=1}^{n} z_{\sigma(i)}) \circ x$. Now let $u_i := z_i$, for all $1 \leq i \leq n$, $u_{n+1} := x$ and $\sigma' \in \mathbb{S}_{n+1}$ with $\sigma'(i) = \sigma(i)$, $\sigma'(n+1) = n+1$. Then, we have
$$a \circ x \subseteq \prod_{i=1}^{n+1} u_i \text{ and } b \circ x \subseteq \prod_{i=1}^{n+1} u_{\sigma'(i)},$$
which implies that $(a \circ x) \overline{\overline{\gamma_{n+1}}} (b \circ x)$. The rest can be proved in an analogous way. ∎

Lemma 2.6.16. *For all $(a, b, x) \in H^3$,*
$$a \gamma_n^* b \Longrightarrow [(a \circ x) \overline{\overline{\gamma_{n+1}^*}} (b \circ x) \text{ and } (x \circ a) \overline{\overline{\gamma_{n+1}^*}} (x \circ b)].$$

Proof. If $a \gamma_n^* b$, then there exists $(z_1, z_2, \ldots, z_m) \in H^m$ such that

$$a = z_1 \gamma_n z_2 \gamma_n \ldots z_{m-1} \gamma_n z_m = b.$$

Now, we have

$$a \circ x = z_1 \circ x \overline{\gamma_{n+1}} z_2 \circ x \ldots z_{m-1} \circ x \overline{\gamma_{n+1}} z_m \circ x = b \circ x.$$

It means that $(a \circ x) \overline{\gamma_{n+1}^*} (b \circ x)$. The rest can be proven similarly. ∎

Let $\gamma_0^* = \emptyset$. We define the γ_n^*-complete hypergroups.

Definition 2.6.17. A hypergroup H is said to be γ_n^*-*complete* if n is the smallest natural number such that $(\gamma_n^*)_H = \gamma_H$ and $(\gamma_n^*)_H \neq (\gamma_{n-1}^*)_H$.

Example 2.6.18. Let H be the following hypergroup:

H	a	b	c	d
a	a, b	c, d	a, b	c, d
b	c, d	a, b	c, d	a, b
c	a, b	c, d	a, b	c, d
d	c, d	a, b	c, d	a, b

Then, H is a γ_3^*-complete hypergroup.

Corollary 2.6.19. *Let H be a commutative hypergroup. Then, H is a γ_n^*-complete hypergroup if and only if H is an n^*-complete hypergroup.*

Proposition 2.6.20. *A hypergroup H is γ_1^*-complete if and only if H is an abelian group.*

Proof. Suppose that H is a γ_1^*-complete hypergroup, so $\gamma_1^* = \gamma$ and $\gamma_2 \subseteq \gamma_1$. Now, for each $x \in z_1 \circ z_2$ and $y \in z_2 \circ z_1$, we have $x \gamma_2 y$, so $x = y$. Therefore, $z_1 \circ z_2 = z_2 \circ z_1$ is singleton, and so H is an abelian group.

Conversely, if H is an abelian group, then $\forall (x, y) \in H^2$; $\prod_{i=1}^{n} z_{\sigma(i)} = \prod_{i=1}^{n} z_i$ and $| \prod_{i=1}^{n} z_i | = 1$. By definition, $x \gamma_n y$ if and only if $x = \prod_{i=1}^{n} z_i$, $y = \prod_{i=1}^{n} z_{\sigma(i)}$. Thus, $x = y$ and $x \gamma_1 y$. ∎

Corollary 2.6.21. *H is a γ_n^*-complete hypergroup if and only if n is the minimum integer such that H/γ_n^* is an abelian group.*

Proposition 2.6.22. *In every hypergroups $(H, \circ)$ the following conditions are equivalent:*

(1) H is γ_2^*-complete hypergroup;

(2) $(H/\gamma_2^*, \otimes)$ is an abelian group;

(3) $(H/\gamma_2^*, \otimes)$ is a commutative hypergroup.

Proposition 2.6.23. *Every finite hypergroup is γ_n^*-complete.*

Proof. Since H is finite, the succession $\gamma_1^* \subseteq \gamma_2^* \subseteq \ldots$ is stationary. Thus, $\exists n \in \mathbb{N}: \quad \gamma_n^* = \gamma$ and $\gamma_n^* \neq \gamma_{n-1}^*$. ∎

Proposition 2.6.24. *If H is a γ_n-complete hypergroup, then there exists $m \leq n$ such that H is γ_m^*-complete.*

Proof. One proves that, if H is γ_n-complete, then $\gamma_n = \gamma$, so $\gamma_n^* = \gamma$ and there exists a $m \leq n$ such that $\gamma_m^* = \gamma$ and $\gamma_{m-1}^* \neq \gamma_m^*$. ∎

Let φ be the canonical projection. Then, we define $D_H = \varphi^{-1}(1_{H/\gamma^*})$.

Theorem 2.6.25. *We have*

(1) *If $(v,w) \in D_H^2$ and $v \, \gamma_n \, w$, then $\gamma = \gamma_{n+1}$;*

(2) *If $(v,w) \in D_H^2$ and $v \, \gamma_n^* \, w$, then $\gamma = \gamma_{n+1}^*$.*

Proof. (1) If $x \, \gamma \, y$, then there exists $(v,w) \in D_H^2$ such that $y \in x \circ v$ and $y \in x \circ w$, by hypothesis $v \, \gamma_n \, w$. Now, using Lemma 2.6.15, $(x \circ v) \, \overline{\overline{\gamma_{n+1}}} \, (x \circ w)$, whence $x \, \gamma_{n+1} \, y$, so $\gamma \subseteq \gamma_{n+1}$.

(2) It follows from (a) and Lemma 2.6.16. ∎

Theorem 2.6.26. *If $(v,w) \in D_H^2$, $v \, \gamma_n^* \, w$ and there exists $(u',w') \in D_H^2$ such that $u' \notin \gamma_{n-1}^* w'$, then H is γ_n^*-complete or γ_{n+1}^*-complete.*

Example 2.6.27. Both of the two possibilities of Theorem 2.6.26 are verifiable, as the following examples shows:

H	a	b	c	d
a	a	b	c	d
b	b	a	d	c
c	c	d	a,b	a,b
d	d	c	a,b	a,b

H'	a	b	c
a	a	b	c
b	b	a,b	c
c	c	c	a,b

Proposition 2.6.28. *If H is a γ_n-complete hypergroup, then for all $(z_1, z_2, \ldots, z_n) \in H^n$ and for all $\sigma \in \mathbb{S}_n$, $\prod_{i=1}^{n} z_{\sigma(i)}$ is a γ-part of H.*

Proof. By using Proposition 2.6.18, the proof is straightforward. ∎

Corollary 2.6.29. *If H is γ_2-complete, then H is a complete hypergroup.*

Corollary 2.6.30. *If H is a γ_2-complete hypergroup, then*

  (1) D_H is the set of identity elements of H.
  (2) H is regular and reversible.

Definition 2.6.31. A K_H *hypergroup* is a hypergroup constructed from a hypergroup $H = (H, \circ)$ and a family $\{A(x)\}_{x \in H}$ of non-empty subsets of H such that

$$\forall (x, y) \in H^2 : \ x \neq y \Longrightarrow A(x) \cap A(y) = \emptyset.$$

Setting $K_H = \bigcup\limits_{x \in H} A(x)$ and defining the following hyperoperation $*$:

$$\forall (a, b) \in K_H^2; \ a \in A(x), \ b \in A(y), \ a * b := \bigcup\limits_{z \in x \circ y} A(z).$$

$(H, \circ)$ is a hypergroup if and only if $(K_H, *)$ is a hypergroup. In this case, K_H is said to be a K_H *hypergroup generated by* H. For all $P \in P^*(H)$, let $K(P) := \bigcup\limits_{x \in P} A(x)$.

Proposition 2.6.32. *$(H, \circ)$ is a hypergroup if and only if $(K_H, *)$ is a hypergroup.*

Proof. It is straightforward. ∎

Theorem 2.6.33. *If P is a γ-part of $(H, \circ)$, then $K(P)$ is γ-part of $(K_H, *)$.*

Proof.[1] Suppose that $(z_1, z_2, \ldots, z_m) \in K_H^m$, and $* \prod\limits_{i=1}^{m} z_i \cap K(P) \neq \emptyset$. Let $u \in * \prod\limits_{i=1}^{m} z_i \cap K(P)$, so there exists $(x_1, x_2, \ldots, x_m) \in H^m$ such that for all $1 \leq i \leq m, \ z_i \in A(x_i)$,

$$u \in \bigcup\limits_{y \in \circ \prod\limits_{i=1}^{m} x_i} A(y) \Longrightarrow \exists \ y_1 \in \circ \prod\limits_{i=1}^{m} x_i : \ u \in A(y_1).$$

---

[1] If $(\circ)$ is a hyperoperation, then $\circ \prod\limits_{i=1}^{m} x_i$ denotes the hyperproduct of the elements x_i by hyperoperation $(\circ)$.

Since $u \in K(P)$, there exists $y_2 \in P$ such that $u \in A(y_2)$. Therefore, $A(y_1) \cap A(y_2) \neq \emptyset$, which implies that $y_1 = y_2 \in \circ \prod_{i=1}^{m} x_i \cap P$. Since P is a γ-part of $(H, \circ)$, for all $\sigma \in \mathbb{S}_m$, $\circ \prod_{i=1}^{m} x_{\sigma(i)} \subseteq P$. Now, for $\sigma \in \mathbb{S}_m$, assume that

$$v \in * \prod_{i=1}^{m} z_{\sigma(i)} = \bigcup_{y \in \circ \prod_{i=1}^{m} x_{\sigma(i)}} A(y).$$

Then, there exists $w \in \circ \prod_{i=1}^{m} x_{\sigma(i)}$ such that $v \in A(w)$. Since $\circ \prod_{i=1}^{m} x_{\sigma(i)} \subseteq P$, we have $A(w) \subseteq K(P)$. Therefore, $v \in K(P)$. ∎

Theorem 2.6.34. *For every $(x, y) \in H^2$ and $(u, v) \in A(x) \times A(y)$, the following conditions are pairwise equivalent:*

(1) $u \ (\gamma_n)_{K_H} \ v$;
(2) $x \ (\gamma_n)_H \ y$;
(3) $A(x) \ \overline{(\gamma_n)_{K_H}} \ A(y)$.

Proof. $(1 \Longrightarrow 2)$ Let $u \ (\gamma_n)_{K_H} \ v$. Then, there exist $(z_1, z_2, \ldots, z_n) \in K_H^n$ and $\sigma \in \mathbb{S}_n$ such that $u \in * \prod_{i=1}^{n} z_i$ and $v \in * \prod_{i=1}^{n} z_{\sigma(i)}$. Now, for every $1 \leq i \leq n$ there exists y_i with $z_i \in A(y_i)$ such that

$$u \in \bigcup_{w \in \circ \prod_{i=1}^{n} y_i} A(w) \quad \text{and} \quad v \in \bigcup_{w' \in \circ \prod_{i=1}^{n} y_{\sigma(i)}} A(w').$$

Thus, there exist $w_1 \in \circ \prod_{i=1}^{n} y_i$ and $w_2 \in \circ \prod_{i=1}^{n} y_{\sigma(i)}$ such that $u \in A(w_1)$, $v \in A(w_2)$. So, $w_1 = x$, $w_2 = y$ and we obtain $x \ (\gamma_n)_H \ y$.

$(2 \Longrightarrow 3)$ Suppose that $x(\gamma_n)_H \ y$. Then, there exist $(z_1, z_2, \ldots, z_n) \in H^n$ and $\sigma \in \mathbb{S}_n$ such that $x \in \circ \prod_{i=1}^{n} z_i$, $y \in \circ \prod_{i=1}^{n} z_{\sigma(i)}$. Now, let $y_i \in A(z_i)$ for $1 \leq i \leq n$. Then,

$$* \prod_{i=1}^{n} y_i = \bigcup_{w \in \circ \prod_{i=1}^{n} z_i} A(w) \quad \text{and} \quad * \prod_{i=1}^{n} y_{\sigma(i)} = \bigcup_{w' \in \circ \prod_{i=1}^{n} z_{\sigma(i)}} A(w').$$

So, $A(x) \subseteq * \prod_{i=1}^{n} y_i$ and $A(y) \subseteq * \prod_{i=1}^{n} y_{\sigma(i)}$. Therefore, we have $A(x) \ \overline{(\gamma_n)_{K_H}} \ A(y)$.

$(3 \Longrightarrow 1)$ It is clear. ∎

Theorem 2.6.35. *For every $(x, y) \in H^2$ and $(u, v) \in A(x) \times A(y)$, the following conditions are pairwise equivalent.*

(1) $u(\gamma_n^*)_{K_H} v$;

(2) $x(\gamma_n^*)_H y$;

(3) $A(x)\,\overline{(\gamma_n^*)_{K_H}}\,A(y)$.

Proof. (1$\Longrightarrow$2) Let $u\,(\gamma_n^*)_{K_H}\,v$. Then, there exist $m \in \mathbb{N}$, $(x_0, x_1, \ldots, x_m) \in K_H^{m+1}$ such that $u = x_0\,(\gamma_n)_{K_H}\,x_1 \ldots x_{m-1}\,(\gamma_n)_{K_H} x_m = v$. For all $0 \le j < m-1$, there exist $(z_1^j, z_2^j, \ldots, z_n^j) \in K_H^n$ and $\sigma^j \in \mathbb{S}_n$ such that $x_j \in *\prod_{i=1}^{n} z_i^j$ and $x_{j+1} \in *\prod_{i=1}^{n} z_{\sigma^j(i)}^j$. Now, for every $1 \le i \le n$, $0 \le j < m-1$, there exists y_i^j such that $z_i^j \in A(y_i^j)$,

$$x_j \in \bigcup_{w \in \circ\, \prod_{i=1}^{n} y_i^j} A(w) \text{ and } x_{j+1} \in \bigcup_{w' \in \circ\, \prod_{i=1}^{n} y_{\sigma^j(i)}^j} A(w').$$

Thus, there exist $w_j \in \circ\prod_{i=1}^{n} y_i^j$ and $w_{j+1} \in \circ\prod_{i=1}^{n} y_{\sigma^j(i)}^j$, where $x_j \in A(w_j)$ and $x_{j+1} \in A(w_{j+1})$ such that $w_0\,(\gamma_n)_H\,w_1 \ldots w_{m-1}\,(\gamma_n)_H\,w_m$. So, $w_0 = x$, $w_m = y$. Therefore, we have $x\,(\gamma_n^*)_H\,y$.

(2$\Longrightarrow$3) Suppose that $x\,(\gamma_n^*)_H\,y$. Then, there exist $m \in \mathbb{N}$ and $(x_0, x_1, \ldots, x_m) \in H^{m+1}$ such that $x = x_0\,(\gamma_n)_H\,x_1 \ldots x_{m-1}\,(\gamma_n)_H\,x_m = y$. For all $0 \le j < m-1$, there exist $(z_1^j, z_2^j, \ldots, z_n^j) \in H^n$ and $\sigma^j \in \mathbb{S}_n$ such that $x_j \in \circ\prod_{i=1}^{n} z_i^j$ and $x_{j+1} \in \circ\prod_{i=1}^{n} z_{\sigma^j(i)}^j$. For all $y_i^j \in A(z_i^j)$, we have

$$*\prod_{i=1}^{n} y_i^j = \bigcup_{w \in \circ\, \prod_{i=1}^{n} z_i^j} A(w) \text{ and } *\prod_{i=1}^{n} y_{\sigma^j(i)}^j = \bigcup_{w' \in \circ\, \prod_{i=1}^{n} z_{\sigma^j(i)}^j} A(w').$$

For all $0 \le j < m-1$, $A(x_j) \subseteq *\prod_{i=1}^{n} y_i^j$ and $A(x_{j+1}) \subseteq *\prod_{i=1}^{n} y_{\sigma(i)}^j$. Now, we have

$$A(x_0)\,\overline{(\gamma_n)_{K_H}}\,A(x_1) \ldots A(x_{m-1})\,\overline{(\gamma_n)_{K_H}}\,A(x_m).$$

Therefore, $A(x)\,\overline{(\gamma_n^*)_{K_H}}\,A(y)$.

(3$\Longrightarrow$1) It is clear. $\blacksquare$

Theorem 2.6.36. *If H is hypergroup and $n \ge 2$, then*

(1) $(\gamma_n)_H = \gamma_H \Longleftrightarrow (\gamma_n)_{K_H} = \gamma_{K_H}$;

(2) $(\gamma_n^*)_H = \gamma_H^* \Longleftrightarrow (\gamma_n^*)_{K_H} = \gamma_{K_H}^*$.

Proof. (1) Suppose that $(\gamma_n)_H = \gamma_H$. It is enough to show $\gamma_{K_H} \subseteq (\gamma_n)_{K_H}$. If $u \ \gamma_{K_H} \ v$, then there exists $(x, y) \in H^2$ such that $u \in A(x)$ and $v \in A(y)$. So, $x \ \gamma_H \ y$ and $x \ (\gamma_n)_H \ y$. Thus, we have $A(x) \ \overline{(\gamma_n)_{K_H}} \ A(y)$, and so $u \ (\gamma_n)_{K_H} \ v$.

Conversely, if $(\gamma_n)_{K_H} = \gamma_{K_H}$, then it is enough to show $\gamma_H \subseteq (\gamma_n)_H$. We have $A(x) \ \overline{(\gamma)_{K_H}} \ A(y)$. Hence, $A(x) \ \overline{(\gamma_n)_{K_H}} \ A(y)$, so we have $x \ (\gamma_n)_H \ y$.

(2) The proof is similar to the proof of part (1). ∎

Corollary 2.6.37. *For $n \geq 2$, H is γ_n^*-complete if and only if K_H is γ_n^*-complete.*

Definition 2.6.38. Let A be a non-empty subset of H. The intersection of the γ-parts of H which contain A is called γ-*closure* of A in H. It will be denoted $C_\gamma(A)$.

Theorem 2.6.39. *Let A be a non-empty subset of H. We pose*

- $G_1(A) := A,$
- $G_{n+1}(A) := \{x \mid \exists p \in \mathbb{N}, \ \exists(h_1, \ldots, h_p) \in H^p, \ \exists \sigma \in \mathbb{S}_p :$
$$x \in \prod_{i=1}^{p} h_{\sigma(i)} \ \text{and} \ \prod_{i=1}^{p} h_i \cap G_n(A) \neq \emptyset\},$$
- $G(A) := \bigcup_{1 \leq n} G_n(A).$

Then, $G(A) = C_\gamma(A)$.

Proof. It is necessary to prove

(1) $G(A)$ is a γ-part of H;
(2) If $A \subseteq B$ and B is a γ-part of H, then $G(A) \subseteq B$.

Therefore,

(1) Let $\prod_{i=1}^{p} x_i \cap G(A) \neq \emptyset$. Then, there exists $n \in \mathbb{N}$ such that $\prod_{i=1}^{p} x_i \cap G_n(A) \neq \emptyset$. For every $\sigma \in \mathbb{S}_p$ and $y \in \prod_{i=1}^{p} x_{\sigma(i)}$ we have $y \in G_{n+1}(A)$ and $\prod_{i=1}^{p} x_{\sigma(i)} \subseteq G(A)$, and so $G(A)$ is a γ-part of H.

(2) We have $A = G_1(A) \subseteq B$. Suppose that B is a γ-part of H and $G_n(A) \subseteq B$. We prove that this implies $G_{n+1}(A) \subseteq B$. For every $z \in G_{n+1}(A)$ there exist $p \in \mathbb{N}$, $(x_1, \ldots, x_p) \in H^p$ and $\sigma \in \mathbb{S}_p$ such that $z \in \prod_{i=1}^{p} x_{\sigma(i)}$, $\prod_{i=1}^{p} x_i \cap G_n(A) \neq \emptyset$. Since $G_n(A) \subseteq B$,

$$\prod_{i=1}^{p} x_i \cap B \neq \emptyset. \text{ Hence, } z \in \prod_{i=1}^{p} x_{\sigma(i)} \subseteq B \text{ and so } G_{n+1}(A) \subseteq B. \blacksquare$$

Lemma 2.6.40. *We have*

(1) $\forall n \geq 2, \forall x \in H, \quad G_n(G_2(x)) = G_{n+1}(x)$.

(2) $x \in G_n(y) \Leftrightarrow y \in G_n(x)$.

Proof. (1) We have

$$G_2(G_2(x)) = \left\{ z \mid \exists p \in \mathbb{N}, \exists (h_1, \ldots, h_p) \in H^p, \exists \sigma \in \mathbb{S}_p : z \in \prod_{i=1}^{p} h_{\sigma(i)}, \right.$$
$$\left. \prod_{i=1}^{p} h_i \cap G_2(x) \neq \emptyset \right\}$$
$$= G_3(x).$$

Now, we proceed by induction. Suppose that $G_{n-1}(G_2(x)) = G_n(x)$. Then,

$$G_n(G_2(x)) = \left\{ z \mid \exists p \in \mathbb{N}, \exists (h_1, \ldots, h_p) \in H^p, \exists \sigma \in \mathbb{S}_p : z \in \prod_{i=1}^{p} h_{\sigma(i)}, \right.$$
$$\left. \prod_{i=1}^{p} h_i \cap G_{n-1}(G_2(x)) \neq \emptyset \right\}$$
$$= \left\{ z \mid \exists p \in \mathbb{N}, \exists (h_1, \ldots, h_p) \in H^p, \exists \sigma \in \mathbb{S}_p : z \in \prod_{i=1}^{p} h_{\sigma(i)}, \right.$$
$$\left. \prod_{i=1}^{p} h_i \cap G_n(x) \neq \emptyset \right\}$$
$$= G_{n+1}(x).$$

(2) We prove by induction. It is clear that $x \in G_2(y) \Leftrightarrow y \in G_2(x)$. Suppose that $x \in G_{n-1}(y) \Leftrightarrow y \in G_{n-1}(x)$. If $x \in G_n(y)$, then there exist $q \in \mathbb{N}$, $(a_1, \ldots, a_q) \in H^q$ and $\sigma \in \mathbb{S}_q$ such that

$$x \in \prod_{i=1}^{q} a_{\sigma(i)} \text{ and } \prod_{i=1}^{q} a_i \cap G_{n-1}(y) \neq \emptyset,$$

by this it follows that there exists $v \in \prod_{i=1}^{q} a_i \cap G_{n-1}(y)$. Therefore, $v \in G_2(x)$ is obtained. From $v \in G_{n-1}(y)$ we have $y \in G_{n-1}(G_2(x)) = G_n(x)$. $\blacksquare$

Theorem 2.6.41. *The relation* $x \, G \, y \Leftrightarrow x \in G(\{y\})$ *is an equivalence relation.*

Proof. We write $C_\gamma(x)$ instead of $C_\gamma(\{x\})$. Clearly, G is reflexive. Now, suppose that xGy and yGz. If P is a γ-part of H and $z \in P$, then $C_\gamma(z) \subseteq P$, $y \in P$ and consequently $x \in C_\gamma(y) \subseteq P$. For this reason $x \in C_\gamma(z)$ that is xGz. The symmetrically of G follows in a direct way from the preceding lemma. ■

Theorem 2.6.42. *For all $x, y \in H$, one gets $x \, G \, y \Leftrightarrow x \, \gamma^* \, y$.*

Proof. Let $x\gamma y$. Then, there exists $n \in \mathbb{N}$ such that $x \, \gamma_n \, y$. So, there exist $(z_1, \ldots, z_n) \in H^n$ and $\sigma \in \mathbb{S}_n$ such that $x \in \prod_{i=1}^{n} z_i$ and $y \in \prod_{i=1}^{n} z_{\sigma(i)}$. We have $\prod_{i=1}^{n} z_i \cap \{x\} \neq \emptyset$, so $y \in G_2(x)$. Hence,

$$x \in G_2(y) \implies xGy \implies \gamma \subseteq G.$$

Since G is an equivalence relation, $\gamma^* \subseteq G$.

Conversely, if xGy, then there exists $n \in \mathbb{N}$ such that $x \in G_{n+1}(y)$, from this it follows that $\exists m \in \mathbb{N}, \exists (z_1^1, \ldots, z_m^1) \in H^m, \exists \sigma^1 \in \mathbb{S}_m$:

$$x \in \prod_{i=1}^{m} z_{\sigma^1(i)}^1 \text{ and } \prod_{i=1}^{m} z_i^1 \cap G_n(y) \neq \emptyset.$$

Thus, $x_1 \in \prod_{i=1}^{m} z_i^1 \cap G_n(y)$. Therefore, $x \, \gamma \, x_1$ and $x_1 \in G_n(y)$ and so there exist $r \in \mathbb{N}, (z_1^2, \ldots, z_r^2) \in H^r, \sigma^2 \in \mathbb{S}_r$ such that

$$x_1 \in \prod_{i=1}^{r} z_{\sigma^2(i)}^2, \prod_{i=1}^{r} z_i^2 \cap G_{n-1}(y) \neq \emptyset \implies \exists x_2 \in \prod_{i=1}^{r} z_i^2 \cap G_{n-1}(y) \implies x_1\gamma x_2.$$

So, as a consequence one obtains

$$\exists x_n \in \prod_{i=1}^{s} z_i^n \cap G_{n-(n-1)}(y) \implies x_n \in G_1(y) = \{y\} \implies x_n = y,$$

and so $x \, \gamma \, x_1 \ldots \gamma \, x_n = y$. Therefore, $G \subseteq \gamma^*$. ■

Theorem 2.6.43. *If B is a non-empty subset of H, then*

$$C_\gamma(B) = \bigcup_{b \in B} C_\gamma(b).$$

Proof. It is clear for every $b \in B$, $C_\gamma(b) \subseteq C_\gamma(B)$, because every γ-part containing B contains $\{b\}$. Therefore, $\bigcup_{b \in B} C_\gamma(b) \subseteq C_\gamma(B)$. In order to prove the converse remember that $C_\gamma(B) = \bigcup_{1 \leq n} G_n(B)$. One clearly has

$$G_1(B) = B = \bigcup_{b \in B} \{b\} = \bigcup_{b \in B} G_1(b).$$

We demonstrate the theorem by induction. Suppose that it is true for n, that is, $G_n(B) \subseteq \bigcup_{b \in B} G_n(b)$ and we prove that $G_{n+1}(B) \subseteq \bigcup_{b \in B} G_{n+1}(b)$. If

$z \in G_{n+1}(B)$, then there exist $q \in \mathbb{N}, (x_1, \ldots, x_q) \in H^q$ and $\sigma \in S_q$ such that

$$z \in \prod_{i=1}^{q} x_{\sigma(i)} \text{ and } \prod_{i=1}^{q} x_i \cap G_n(B) \neq \emptyset,$$

by the hypothesis induction $\prod_{i=1}^{q} x_i \cap \left(\bigcup_{b \in B} G_n(b) \right) \neq \emptyset$. Hence, there exists $b' \in B$ such that $\prod_{i=1}^{q} x_i \cap G_n(b') \neq \emptyset$. Since $z \in \prod_{i=1}^{q} x_{\sigma(i)}$, one gets $z \in G_{n+1}(b')$ and so one has prove $G_{n+1}(B) \subseteq \bigcup_{b \in B} G_{n+1}(b)$. Therefore, $C_\gamma(B) \subseteq \bigcup_{b \in B} C_\gamma(b)$. ∎

Theorem 2.6.44. *If $A \in \mathcal{P}^*(H)$, one has $D_H \circ A = A \circ D_H = C_\gamma(A)$.*

Proof. It is straightforward. ∎

Corollary 2.6.45. *Let $A \in \mathcal{P}^*(H)$. Then, A is a γ-part of H if and only if $A \circ D_H = A$.*

Proof. We have $C_\gamma(A) = A \circ D_H = A$. ∎

Corollary 2.6.46. *If A is a γ-part of H, then for every $B \in \mathcal{P}^*(H)$, $A \circ B$ and $B \circ A$ are γ-parts of H.*

Proof. We have: $C_\gamma(A \circ B) = A \circ B \circ D_H = A \circ D_H \circ B = C_\gamma(A) \circ B = A \circ B$. ∎

2.7 Join spaces

Join spaces were introduced by W. Prenowitz [118; 120; 121; 122] to provide a common algebraic framework in which classical geometries could be axiomatized and studied. The underlying algebraic structure used was a hypergroup. Then this concept applied by him and J. Jantosciak both in Euclidian and in non Euclidian geometry [124]. Using this notion, several branches of non Euclidian geometry were rebuilt: descriptive geometry, projective geometry and spherical geometry. Then, several important examples of join spaces have been constructed in connection with binary relations, graphs, lattices, fuzzy sets and rough sets. In this section, we study the concept of join space. The main references for this section are [56; 86].

In order to define a join space, we recall the following notation: If a, b are elements of a hypergroupoid $(H, \circ)$, then we denote $a/b = \{x \in H \mid a \in x \circ b\}$. Moreover, by A/B we intend the set $\bigcup\limits_{a \in A, b \in B} a/b$.

Definition 2.7.1. A commutative hypergroup $(H, \circ)$ is called a *join space* if the following condition holds for all elements a, b, c, d of H:

$$a/b \cap c/d \neq \emptyset \implies a \circ d \cap b \circ c \neq \emptyset \quad \text{(transposition axiom)}.$$

Elements of H are called *points* and are denoted by $a, b, c, \ldots$. Sets of points are denoted by $A, B, C, \cdots$. The elementary algebra of join space theory for sets of points include the result

  (1) $(A/B)/C = A/(B \circ C)$;
  (2) $A \neq \emptyset$ implies $B \subseteq A/(A/B)$;
  (3) $A \circ (B/C) \subseteq (A \circ B)/C$;
  (4) $A/(B/C) \subseteq (A \circ C)/B$.

A set of points M is said to be *linear* if it is closed under join and extension ($a, b \in M$ imply $a \circ b \subseteq M$ and $a/b \subseteq M$). If M is linear, then $M = M \circ M = M/M$. The linear sets are the closed subhypergroups of H. For a set of points A, the intersection of all linear sets containing A (the least linear set containing A) is denoted by $< A >$ and is called the linear space *spanned* or *generated* by A. We use $< A, B >$ for $< A \cup B >$. Important results concerning linear sets are a formula for the linear span of two intersecting linear sets and a weak modularity property

  (1) M and N linear and $M \cap N \neq \emptyset$ imply $< M, N >= M/N$;
  (2) L, M and N linear, $L \cap M \neq \emptyset$ and $L \subseteq N$ imply $< L, M > \cap N =< L, M \cap N >$.

Both of these results depend on the transposition axiom.

A point e is said to be a *(scalar) identity* if $e \circ a = a$ for every point a. If H has an identity, it is unique. In a join space H with identity e the following hold:

  (1) For each a there exists a unique a^{-1} such that $e \in a \circ a^{-1}$;
  (2) $a^{-1} = e/a$;
  (3) $a/b = a \circ b^{-1}$;
  (4) $e \in M$ for any non-empty linear set M.

The transposition axiom is needed here to prove the uniqueness of the inverse point and that an extension reduce to a join. Join spaces with identity have been studied also in [112] by the name of canonical hypergroups.

Some important examples of join spaces were presented in Section 2.2 (see Examples 2.2.3. (9), (10), (11), (12), (14)). We give here some other examples.

Example 2.7.2.

(1) Let $(L, \wedge, \vee)$ be a distributive lattice. If for all a, b of L we define

$$a \circ b = \{x \in L \mid x = (a \wedge b) \vee (a \wedge x) \vee (b \wedge x)\},$$

then $(L, \circ)$ is a non geometrical join space in which every element is an identity. The above hyperoperation can be considered in a more general context, that one of a median semilattice. A *median semi-lattice* is a meet semilattice $(S, \wedge)$, such that the following conditions hold:

- Every principal ideal is a distributive lattice;
- Any three elements of S have an upper bound whenever each pair of them has an upper bound.

(2) Let V be a vector space over an ordered field F. If for all a, b of V we define

$$a \circ b = \{\lambda a + \mu b \mid \lambda > 0, \mu > 0, \ \lambda + \mu = 1\},$$

then $(V, \circ)$ is a join space, called an *affine join space* over F.

(3) Let $G = (V, E)$ be a connected simple graph. We say that a subset A of V is convex if for all different elements a, b of A, we have that A contains all points on all geodetics from a to b. Denote by (a, b) the least convex set containing $\{a, b\}$. A convex set P is called *prime* if $V \setminus P$ is convex. Finally, G is called a *strong prime convex intersection graph* if:

- For any convex set A and any point x, which does not belong to A, there exists a prime convex set P, such that $A \subseteq P$, $x \in V \setminus P$;
- For any (a, b), (c, d) such that $(a, b) \cap (c, d) = \emptyset$, there exists a convex prime set P such that $(a, b) \subseteq P$ and $(c, d) \subseteq V \setminus P$.

If G satisfies the above two conditions and for all different elements a, b of V we define $a \circ b = (a, b)$ and $a \circ a = a$, then $(V, \circ)$ is a join space.

(4) Denote by $)a, b($ an open real interval. We define the following hyperoperation on the Cartesian plane $\mathbb{R}^2$: for all different elements $(x_1, x_2), (y_1, y_2)$ of $\mathbb{R}^2$, we have $(x_1, x_2) \circ (y_1, y_2) = \{(z_1, z_2) \mid z_1 \in)x_1, x_2($ and $z_2 \in)x_2, y_2(\}$ and for all element (x_1, x_2) of $\mathbb{R}^2$, we

have $(x_1, x_2) \circ (x_1, x_2) = (x_1, x_2)$. Then, $(\mathbb{R}^2, \circ)$ is a geometric join space, not provided with identity elements.

(5) Let $G = (V, E)$ be a connected simple graph. We define the following hyperoperation on V: for all different elements x, y of V, we have $x \circ x = x$ and $x \circ y$ is the set of all points $z \in V$, which belong to some paths $\gamma : x - y$. Then, $(V, \circ)$ is a non-geometric join space in which every element is an identity.

Three classical types of geometry are readily formulated as join spaces.

Definition 2.7.3. (1) A *descriptive join space* or *ordered join geometry* is a join space that satisfies the axioms

$a \circ a = a/a = a$;

a, b, c distinct and $c \in \{a, b\}$ imply $c \in a \circ b$, $b \in a \circ c$ or $a \in b \circ c$.

(2) A *spherical join space* is a join space with identity e that satisfies the axioms

$a \circ a = a$;

$a/a = \{e, a, a^{-1}\}$.

(3) A *projective join space* is a join space with identity e that satisfies the axiom

$a \circ a = a/a \subseteq \{e, a\}$.

Proposition 2.7.4. *If H is one of the classical join spaces, then the "line" spanned by points a and b, i.e., $< a, b >$ is given by*

(1) $a \cup b \cup a \circ b \cup a/b \cup b/a$, for H descriptive;

(2) $e \cup a \cup b \cup a^{-1} \cup b^{-1} \cup a \circ b \cup a \circ b^{-1} \cup a^{-1} \circ b \cup a^{-1} \circ b^{-1}$, for H spherical;

(3) $e \cup a \cup b \cup a \circ b$, for H projective.

Proof. It is straightforward. ■

Let M be a non-empty linear subset of the join space H. For any point a the *coset* of M in H containing a is denoted by $(a)_M$ and is given by $(a)_M = (a \circ M)/M$. The family $(H : M) = \{(a)_M \mid a \in H\}$ of all cosets of M in H is a partition of H. For the set of points A let

$$(A)_M = (A \circ M)/M = \cup\{(a)_M \mid a \in A\}.$$

Then, the join cosets $(a)_M$ and $(b)_M$ satisfies

$$(a)_M \circ (b)_M \subseteq (a \circ b)_M,$$

so that the equivalence relation on H corresponding to the partition of H into cosets of M is a regular equivalence relation.

Proposition 2.7.5. *The family $(H : M)$ of cosets M in H under the join operation*

$$(a)_M \star (b)_M = \{(x)_M \mid x \in (a)_M \circ (b)_M\}$$

is a join space with identity $M = (m)_M$, where $m \in M$.

Proof. Note that $(a)_M^{-1} = M/a$ and that $a \circ b \cap M \neq \emptyset$ if and only if $(a)_M^{-1} = (b)_M$. ∎

$(H : M)$ under $\star$ is known as the *factor join space H modulo M*.

Analogues of the three classical isomorphism theorems of group theory hold for join spaces. Let $(H, \circ)$ and $(K, \cdot)$ be join spaces. A mapping $\varphi : H \longrightarrow K$ is said to be a *good homomorphism* if $\varphi(a \circ b) = \varphi(a) \cdot \varphi(b)$. If φ is also one to one and onto, then φ is said to be an *isomorphism* and the notation $H \cong K$ is used. If φ is a good homomorphism and K has identity e, then $\{x \in H \mid \varphi(x) = e\}$ is called the *kernel* of φ and is denoted by $ker\varphi$.

Theorem 2.7.6. (First Isomorphism Theorem) *Let H and K be join spaces. Let K has identity. Let φ be a good homomorphism of H onto K. Then, $ker\varphi$ is a linear subset of H and $(H : ker\varphi) \cong K$.*

Theorem 2.7.7. (Second Isomorphism Theorem) *Let H be a join space. Let M and N be linear subsets such that $M \cap N \neq \emptyset$. Then,*

$$(< M, N >: M) \cong (N : M \cap N).$$

Theorem 2.7.8. (Third Isomorphism Theorem) *Let H be a join space. Let M and N be linear subsets such that $\emptyset \neq N \subseteq M$. Then, $(M : N)$ is a linear subset of $(H : N)$ and $(H : M) \cong ((H : N) : (M : N))$.*

A version of Jordan-Hölder theorem also holds for join spaces.

Theorem 2.7.9. (Jordan-Hölder Theorem) *Let H be a join space. Let M and N be linear subsets such that $\emptyset \neq N \subseteq M$. Suppose that*

$$N = A_0 \subseteq \cdots \subseteq A_m = M \text{ and } N = B_0 \subseteq \cdots \subseteq B_n = M$$

where each A_i and each B_j are linear and for $i = 1, \cdots, m$ and for $j = 1, \cdots, n$ that A_{i-1} and B_{j-1} are maximal proper linear subsets of A_i and

B_j respectively. Then, $m = n$ and there exists a one to one corresponding between the families of factor spaces

$$\{(A_i : A_{i-1}) \mid i = 1, \cdots, m\} \text{ and } \{(B_j : B_{j-1}) \mid j = 1, \cdots, n\}$$

such that the correspondents are isomorphic.

Proof. The chains $A_0 \subseteq \cdots \subseteq A_m$ and $B_0 \subseteq \cdots \subseteq B_n$ are refined respectively by

$$A_{i,j} = < A_{i-1}, A_i \cap B_j > \text{ for } i = 1, \cdots, m \text{ and } j = 0, \cdots, n$$

and

$$B_{j,i} = < B_{j-1}, B_j \cap A_i > \text{ for } i = 0, \cdots, m \text{ and } j = 1, \cdots, n.$$

Thus,

$$A_0 = A_{1,0} \subseteq \cdots \subseteq A_{1,n} = A_1 = A_{2,0} \subseteq \cdots \subseteq A_{m,n} = A_m$$

and

$$B_0 = B_{1,0} \subseteq \cdots \subseteq B_{1,m} = B_1 = B_{2,0} \subseteq \cdots \subseteq B_{n,m} = B_n.$$

Next, by the second isomorphism theorem followed by the weak modularity property

$$
\begin{aligned}
(A_{i,j} : A_{i,j-1}) &= (< A_{i-1}, A_i \cap B_j > :< A_{i-1}, A_i \cap B_{j-1} >) \\
&= (<< A_{i-1}, A_i \cap B_{j-1} >, A_i \cap B_j > :< A_{i-1}, A_i \cap B_{j-1} >) \\
&\cong (A_i \cap B_j :< A_{i-1}, A_i \cap B_{j-1} > \cap (A_i \cap B_j)) \\
&= (A_i \cap B_j :< A_{i-1} \cap (A_i \cap B_j), A_i \cap B_{j-1} >) \\
&= (A_i \cap B_j :< A_{i-1} \cap B_j, A_i \cap B_{j-1} >).
\end{aligned}
$$

Similarly, we have

$$(B_{j,i} : B_{j,i-1}) \cong (B_j \cap A_i :< B_{j-1} \cap A_i, B_j \cap A_{i-1} >).$$

Hence,

$$(A_{i,j} : A_{i,j-1}) \cong (B_{j,i} : B_{j,i-1}) \text{ for } i = 1, \cdots, m \text{ and } j = 1, \cdots, n.$$

Therefore, given the maximality condition on A_{i-1} and B_{j-1} in A_i and B_j respectively, the conclusion of the theorem readily follows. ∎

If N is a closed subhypergroup of a join space H and $\{x, y\} \subseteq H$, then we define the following binary relation: $x J_N y$ if $x \circ N \cap y \circ N \neq \emptyset$.

Theorem 2.7.10. J_N *is an equivalence relation on H and the equivalence class of an element a is $J_N(a) = (a \circ N)/N$. In particular, $J_N(a) = N$ for all $a \in N$.*

Proof. Clearly, J_N is reflexive and symmetric. Now, suppose that $a \circ N \cap b \circ N \neq \emptyset$ and $b \circ N \cap c \circ N \neq \emptyset$. It follows that $b \in (a \circ N)/N \cap (c \circ N)/N$ and since $(H, \circ)$ is a join space, we obtain $a \circ N \cap c \circ N \neq \emptyset$, which means that $a J_N c$. Hence, J_N is also transitive, and so it is an equivalence relation on H. We check now that for all $a \in H$ we have $J_N(a) = (a \circ N)/N$. If $d \in J_N(a)$, then $d \circ N \cap a \circ N \neq \emptyset$. Hence, there exist $v \in a \circ N$ and $m \in N$ such that $v \in d \circ m$, whence it follows that $d \in v/m \subseteq (a \circ N)/N$. We obtain $J_N(a) \subseteq (a \circ N)/N$. Now, let $y \in (a \circ N)/N$. Then, there exist $u \in a \circ N$ and $m \in N$, such that $u \in y \circ m$, whence $y \circ N \cap a \circ N \neq \emptyset$, which means that $y \in J_N(a)$ and so, $(a \circ N)/N \subseteq J_N(a)$. Clearly, if $a \in N$, then $J_N(a) = N$, since N is closed. ∎

Canonical hypergroups are a particular case of join spaces. The structure of canonical hypergroups was individualized for the first time by M. Krasner as the additive structure of hyperfields. In 1970, J. Mittas was the first who studied them independently from the other operations. In 1973, P. Corsini analyzed the sd-hypergroups, which are a particular type of canonical hypergroups and in 1975 Roth used canonical hypergroups in the character theory of finite groups. W. Prenowitz and J. Jantosciak emphasized the role of canonical hypergroups in geometry, while J.R. Mc-Mullen and J.F. Price underlined the role of a generalization of canonical hypergroups in harmonic analysis and particle physics. Some connected hyperstructures with canonical hypergroups were introduced and analyzed by P. Corsini, P. Bonansinga, K. Serafimidis, M. Kostantinidou, J. Mittas, De Salvo. We mention here some of them: strongly canonical, i.p.s. hypergroups, quasi-canonical hypergroups (also called polygroups), feebly (quasi)canonical hypergroups.

Let us see now what a canonical hypergroup is.

Definition 2.7.11. We say that a hypergroup $(H, \circ)$ is *canonical* if

(1) it is commutative,
(2) it has a scalar identity (also called scalar unit), which means that
$$\exists e \in H, \ \forall x \in H, \ x \circ e = e \circ x = x,$$
(3) every element has a unique inverse, which means that for all $x \in H$, there exists a unique $x^{-1} \in H$, such that $e \in x \circ x^{-1} \cap x^{-1} \circ x$,
(4) it is reversible, which means that if $x \in y \circ z$, then there exist the inverses y^{-1} of y and z^{-1} of z, such that $z \in y^{-1} \circ x$ and $y \in x \circ z^{-1}$.

Clearly, the identity of a canonical hypergroup is unique. Indeed, if e is a scalar identity and e' is an identity of a canonical hypergroup $(H, \circ)$, then we have $e \in e \circ e' = \{e'\}$.

Some interesting examples of a canonical hypergroup is the following ones (see [9]).

Example 2.7.12.

(1) Let $C(n) = \{e_0, e_1, \cdots, e_{k(n)}\}$, where $k(n) = n/2$ if n is an even natural number and $k(n) = (n-1)/2$ if n is an odd natural number. For all e_s, e_t of $C(n)$, define $e_s \circ e_t = \{e_p, e_v\}$, where $p = \min\{s + t, n - (s+t)\}$, $v = |s-t|$. Then $(C(n), \circ)$ is a canonical hypergroup.

(2) Let (S, T) be a projective geometry, i.e., a system involving a set S of elements called *points* and a set T of sets of points called *lines*, which satisfies the following postulates:

  • Any lines contains at least three points;
  • Two distinct points a, b are contained in a unique line, that we shall denote by $L(a, b)$;
  • If a, b, c, d are distinct points and $L(a, b) \cap L(c, d) \neq \emptyset$, then $L(a, c) \cap L(b, d) \neq \emptyset$.

Let e be an element which does not belong to S and let $S' = S \cup \{e\}$. We define the following hyperoperation on S':

  • For all different points a, b of S, we consider $a \circ b = L(a, b) \setminus \{a, b\}$;
  • If $a \in S$ and any line contains exactly three points, let $a \circ a = \{e\}$, otherwise $a \circ a = \{a, e\}$;
  • For all $a \in S'$, we have $e \circ a = a \circ e = a$.

Then $(S', \circ)$ is a canonical hypergroup.

In what follows, we present some basic results of canonical hypergroups.

Theorem 2.7.13. *If $(H, \circ)$ is a canonical hypergroup, then the following implication holds for all x, y, z, t of H:*

$$x \circ y \cap z \circ t \neq \emptyset \implies x \circ z^{-1} \cap t \circ y^{-1} \neq \emptyset.$$

Proof. Let $u \in x \circ y \cap z \circ t$. Since H is reversible, we obtain $u^{-1} \in z^{-1} \circ t^{-1}$, whence $u \circ u^{-1} \subseteq x \circ y \circ z^{-1} \circ t^{-1}$. If e is an identity of H, we obtain $e \in (x \circ z^{-1}) \circ (t \circ y^{-1})^{-1}$. Hence there exists an element $v \in x \circ z^{-1} \cap t \circ y^{-1}$. ∎

Theorem 2.7.14. *A commutative hypergroup is canonical if and only if it is a join space with a scalar identity.*

Proof. Suppose that $(H, \circ)$ is a canonical hypergroup. For all a, b of H we have $a/b = a \circ b^{-1}$. Then the implication $\Longrightarrow$ follows by the above theorem.

Conversely, let us check that the inverse of an element is unique. Let e be the scalar identity. If $e \in a \circ b \cap a \circ c$, then $a \in e/b \cap e/c$, whence it follows that $e \circ c \cap e \circ b \neq \emptyset$, hence $b = c = a^{-1}$. Let us check now the reversibility of H. We have $a \in b \circ c$ if and only if $b \in a/c$. From $e \in b \circ b^{-1}$ we obtain $b \in e/b^{-1}$, hence $a \circ b^{-1} \cap e \circ c \neq \emptyset$, which means that $c \in a \circ b^{-1}$. Therefore, H is canonical. ∎

Theorem 2.7.15. *If $(H, \circ)$ is a join space and N is a closed subhypergroup of H, then the quotient $(H/J_N, \otimes)$ is a canonical hypergroup, where for all $\overline{a}, \overline{b}$ of H/J_N, we have $\overline{a} \otimes \overline{b} = \{\overline{c} \mid c \in a \circ b\}$.*

Proof. First, we check that the hyperoperation $\otimes$ is well defined. In other words, we have to check that if $a_1 J_N a_2$ and $x \in H$, then for all $z \in a_1 \circ x$, there exists $w \in a_2 \circ x$, such that $z J_N w$. Indeed, from $a_1 J_N a_2$, it follows there exist m, n of N and v of H, such that $v \in a_1 \circ m \cap a_2 \circ n$. If $z \in a_1 \circ x$, then we have $a_1 \in z/x \cap v/m$, hence $z \circ N \cap v \circ x \neq \emptyset$, whence $z \circ N \cap N \circ a_2 \circ x \neq \emptyset$. It follows that there exists $w \in a_2 \circ x$, such that $z J_N w$. Therefore the hyperoperation $\otimes$ is well defined. Since $(H, \circ)$ is a join space, it follows that $(H/J_N, \otimes)$ is a join space, too. Moreover, notice that N is a scalar identity for $(H/J_N, \otimes)$, and according to the above theorem, we obtain that $(H/J_N, \otimes)$ is a canonical hypergroup. ∎

Chapter 3

Polygroups

3.1 Definition and examples of polygroups

Quasicanonical hypergroups were introduced by P. Corsini and later, they were studied by P. Bonansinga and Ch.G. Massouros. They satisfy all the conditions of canonical hypergroups, except the commutativity. Later, S.D. Comer introduced this class of hypergroups independently, using the name of polygroups. He emphasized the importance of polygroups, by analyzing them in connections to graphs, relations, Boolean and cylindric algebras. Another connection between polygroups and artificial intelligence was considered and analyzed by G. Ligozat. Some of these results are exposed in [31]. The double cosets hypergroups are particular quasicanonical hypergroups and they were analyzed by K. Drbohlav, D.K. Harrison and S.D. Comer.

Definition 3.1.1. A *polygroup* is a system $\wp = <P, ., e, ^{-1}>$, where $e \in P$, $^{-1}$ is a unitary operation on P, $\cdot$ maps $P \times P$ into the non-empty subsets of P, and the following axioms hold for all $x, y, z \in P$:

(P$_1$) $(x \cdot y) \cdot z = x \cdot (y \cdot z)$,
(P$_2$) $e \cdot x = x \cdot e = x$,
(P$_3$) $x \in y \cdot z$ implies $y \in x \cdot z^{-1}$ and $z \in y^{-1} \cdot x$.

The following elementary facts about polygroups follow easily from the axioms: $e \in x \cdot x^{-1} \cap x^{-1} \cdot x$, $e^{-1} = e$, $(x^{-1})^{-1} = x$, and $(x \cdot y)^{-1} = y^{-1} \cdot x^{-1}$, where $A^{-1} = \{a^{-1} | \ a \in A\}$.

A polygroup in which every element has order 2 (i.e., $x^{-1} = x$ for all x) is called *symmetric*. As in group theory it can be shown that a symmetric polygroup is commutative.

Example 3.1.2.

(1) *Double coset algebra.* Suppose that H is a subgroup of a group G. Define a system $G//H =< \{HgH \mid g \in G\}, *, H, ^{-I} >$, where $(HgH)^{-I} = Hg^{-1}H$ and

$$(Hg_1H) * (Hg_2H) = \{Hg_1hg_2H \mid h \in H\}.$$

The algebra of double cosets $G//H$ is a polygroup introduced in (Dresher and Ore [69]).

(2) *Prenowitz algebras.* Suppose G is a projective geometry with a set P of points and suppose, for $p \neq q$, $\overline{pq}$ denoted the set of all points on the unique line through p and q. Choose an object $I \notin P$ and form the system

$$P_G =< P \cup \{I\}, \cdot, I, ^{-1} >$$

where $x^{-1} = x$ and $I \cdot x = x \cdot I = x$ for all $x \in P \cup \{I\}$ and for $p, q \in P$,

$$p \cdot q = \begin{cases} \overline{pq} \setminus \{p, q\} & \text{if } p \neq q \\ \{p, I\} & \text{if } p = q. \end{cases}$$

P_G is a polygroup (Prenowitz [119]).

(3) *Conjugacy class polygroups.* In dealing with a symmetry group two symmetric operations belong to the same class if they present the same map with respect to (possibly) different coordinate systems where one coordinate system is converted into the other by a member of the group. In the language of group theory this means the elements a, b in a symmetric group G belong to the same class if there exists a $g \in G$ such that $a = gbg^{-1}$, i.e., a and b are conjugate. The collection of all conjugacy classes of a group G is denoted by $\overline{G}$ and the system $< \overline{G}, *, \{e\}, ^{-1} >$ is a polygroup where e is the identity of G and the product $A * B$ of conjugacy classes A and B consists of all conjugacy classes contained in the elementwise product AB. This hypergroup was recognized by Campaigne [8] and by Diatzman [68].

Now, we illustrate constructions using the dihedral group D_4. This group is generated by a counter-clockwise rotation r of $90°$ and a horizontal reflection h. The group consists of the following 8 symmetries:

$$\{1 = r^0, r, r^2 = s, r^3 = t, h, hr = d, hr^2 = v, hr^3 = f\}.$$

The dihedral groups occur frequently in art and nature. Many of the decorative designs used on floor coverings, pottery, and buildings have one of the dihedral groups as a group of symmetry. In the case of D_4 there are five conjugacy classes: $\{1\}, \{s\}, \{r, t\}, \{d, f\}$ and $\{h, v\}$. Let us denote these classes by $C_1, \ldots, C_5$ respectively. Then, the polygroup $\overline{D}_4$ is

$*$	C_1	C_2	C_3	C_4	C_5
C_1	C_1	C_2	C_3	C_4	C_5
C_2	C_2	C_1	C_3	C_4	C_5
C_3	C_3	C_3	C_1, C_2	C_5	C_4
C_4	C_4	C_4	C_5	C_1, C_2	C_3
C_5	C_5	C_5	C_4	C_3	C_1, C_2

As a sample of how to calculate the table entries consider $C_3 \cdot C_3$. To determine this product compute the elementwise product of the conjugacy classes $\{r, t\}\{r, t\} = \{s, 1\} = C_1 \cup C_2$. Thus, $C_3 \cdot C_3$ consists of the two conjugacy classes C_1, C_2.

(4) *Character polygroups.* Closely related to the conjugacy classes of a finite group are its characters. Let $\hat{G} = \{\chi_1, \chi_2, \ldots, \chi_k\}$ be the collection of irreducible characters of a finite group G where χ_1 is trivial character. The character polygroup $\hat{G}$ of G is the system $< \hat{G}, *, \chi_1, ^{-1} >$ where the product $\chi_i * \chi_j$ is the set of irreducible components in the elementwise product $\chi_i \chi_j$. The system $\hat{G}$ was investigated by R. Roth [129] who consider a duality between $\hat{G}$ and $\overline{G}$.

Before calculating $\hat{D}_4$ we need to know the five irreducible characters of the dihedral group D_4. These are given by the following character table. (since characters are constant on conjugacy classes it is usual to list only the conjugacy classes across the top of the table.)

	C_1	C_2	C_3	C_4	C_5
$\chi_1:$	1	1	1	1	1
$\chi_2:$	1	1	-1	1	-1
$\chi_3:$	1	1	-1	-1	1
$\chi_4:$	1	1	1	-1	-1
$\chi_5:$	2	-2	0	0	0

We illustrate the calculation of the polygroup product of two characters by considering $\chi_5 * \chi_5$. The pointwise product of χ_5 with itself yields the following (non-irreducible) character:

$$\frac{\mathcal{C}_1\ \mathcal{C}_2\ \mathcal{C}_3\ \mathcal{C}_4\ \mathcal{C}_5}{\chi_5\chi_5 : 4\ \ 4\ \ 0\ \ 0\ \ 0}$$

This character can be written as a sum of irreducible characters in exactly one way: $\chi_5\chi_5 = \chi_1 + \chi_2 + \chi_3 + \chi_4$. This is indicated by the entry in the lower right hand corner of the polygroup table for $\hat{D}_4$. In general the polygroup product of two characters $\chi_i * \chi_j$ tells which irreducible characters are in the product $\chi_i\chi_j$, but not the multiplicity. Using i in place of the character χ_i the polygroup $\hat{D}_4$ is

	1	2	3	4	5
1	1	2	3	4	5
2	2	1	4	3	5
3	3	4	1	2	5
4	4	3	2	1	5
5	5	5	5	5	$\{1,2,3,4\}.$

3.2   Extension of polygroups by polygroups

Let $< P_1, \cdot, e_1, ^{-1} >$ and $< P_2, *, e_2, ^{-I} >$ be two polygroups. Then, on $P_1 \times P_2$ we can define a hyperproduct as follows: $(x_1, y_1) \circ (x_2, y_2) = \{(x, y) \mid x \in x_1 x_2, \ y \in y_1 * y_2\}$. We recall this the *direct hyperproduct* of P_1 and P_2. Clearly, $P_1 \times P_2$ equipped with the usual direct hyperproduct becomes a polygroup.

In [20], extensions of polygroups by polygroups were introduced in the following way.

Suppose that $\mathcal{A} =< A, \cdot, e, ^{-1} >$ and $\mathcal{B} =< B, \cdot, e, ^{-1} >$ are two polygroups whose elements have been renamed so that $A \cap B = \{e\}$. A new system $\mathcal{A}[\mathcal{B}] =< M, *, e, ^I >$ called the *extension* of $\mathcal{A}$ by $\mathcal{B}$ is formed in the following way: Set $M = A \cup B$ and let $e^I = e$, $x^I = x^{-1}$, $e * x = x * e = x$ for all $x \in M$, and for all $x, y \in M \setminus \{e\}$

$$x * y = \begin{cases} x \cdot y & \text{if } x, y \in A \\ x & \text{if } x \in B, \ y \in A \\ y & \text{if } x \in A, \ y \in B \\ x \cdot y & \text{if } x, y \in B, \ y \neq x^{-1} \\ x \cdot y \cup A & \text{if } x, y \in B, \ y = x^{-1}. \end{cases}$$

In this case $\mathcal{A}[\mathcal{B}]$ is a polygroup which is called the extension of $\mathcal{A}$ by $\mathcal{B}$.

In the last case, e occurs in both $x \cdot y$ and A. If $A = \{e, a_1, a_2, \ldots\}$ and $B = \{e, b_1, b_2, \ldots\}$, the table for $*$ in $\mathcal{A}[\mathcal{B}]$ has the form

	e	a_1	a_2	$\ldots$	b_1	b_2	$\ldots$
e	e	a_1	a_2	$\ldots$	b_1	b_2	$\ldots$
a_1	a_1	$a_1 a_1$	$a_1 a_2$	$\ldots$	b_1	b_2	$\ldots$
a_2	a_2	$a_2 a_1$	$a_2 a_2$	$\ldots$	b_1	b_2	$\ldots$
$\vdots$	$\vdots$	$\vdots$	$\vdots$	$\vdots$	$\vdots$	$\vdots$	$\vdots$
b_1	b_1	b_1	b_1	$\ldots$	$b_1 * b_1$	$b_1 * b_2$	$\ldots$
b_2	b_2	b_2	b_2	$\ldots$	$b_2 * b_1$	$b_2 * b_2$	$\ldots$
$\vdots$	$\vdots$	$\vdots$	$\vdots$	$\vdots$	$\vdots$	$\vdots$	$\vdots$

Several special cases of the algebra $\mathcal{A}[\mathcal{B}]$ are useful. Before describing them we need to assign names to the two 2-elements polygroups. Let **2** denotes the group $\mathbb{Z}_2$ and let **3** denotes the polygroup $\mathbb{S}_3// < (1\ 2) > \cong \mathbb{Z}_3/\theta$, where θ is the special conjugation with blocks $\{0\}, \{1, 2\}$. The multiplication table for **3** is

	0	1
0	0	1
1	1	$\{0, 1\}$

The names **2** and **3** are suggested by the color schemes that represent the algebras (see Chapter 5).

Example 3.2.1. (Adjoining a new identity element). The system $\mathbf{3}[\mathcal{M}]$ is the result of adding a "new" identity to the polygroup $\mathcal{M}$. The system $\mathbf{2}[\mathcal{M}]$ is almost as good. For example, suppose that $\mathcal{R}$ is the system with table

	0	1	2
0	0	1	2
1	1	$\{0, 2\}$	$\{1, 2\}$
2	2	$\{1, 2\}$	$\{0, 1\}$

Then,

	0	a	1	2
0	0	a	1	2
a	a	$\{0, a\}$	1	2
1	1	1	$\{0, a, 2\}$	$\{1, 2\}$
2	2	2	$\{1, 2\}$	$\{0, a, 1\}$

$\mathbf{3}[\mathcal{R}]$

	0	a	1	2
0	0	a	1	2
a	a	0	1	2
1	1	1	$\{0, a, 2\}$	$\{1, 2$
2	2	2	$\{1, 2\}$	$\{0, a, 1\}$

$\mathbf{2}[\mathcal{R}]$

The element "a" acts like the "old" identity on $\mathcal{R}$.

Example 3.2.2. The tables for $\mathcal{R}[2]$ and $\mathcal{R}[3]$ are given below:

	0	1	2	a
0	0	1	2	a
1	1	$\{0,2\}$	$\{1,2\}$	a
2	2	$\{1,2\}$	$\{0,1\}$	a
a	a	a	a	$\{0,1,2\}$

$$\mathcal{R}[2]$$

	0	1	2	a
0	0	1	2	a
1	1	$\{0,2\}$	$\{1,2\}$	a
2	2	$\{1,2\}$	$\{0,1\}$	a
a	a	a	a	$\{0,1,2,a\}$

$$\mathcal{R}[3]$$

Example 3.2.3. As an example of $\mathcal{A}[\mathcal{B}]$, where neither $\mathcal{A}$ nor $\mathcal{B}$ are minimal, we consider $\mathcal{R}[\mathcal{R}]$ whose table is given below:

	0	1	2	a	b
0	0	1	2	a	b
1	1	$\{0,2\}$	$\{1,2\}$	a	b
2	2	$\{1,2\}$	$\{0,1\}$	a	b
a	a	a	a	$\{0,1,2,b\}$	$\{a,b\}$
b	b	b	b	a,b	$\{0,1,2,a\}$

We finish this section by showing that the extension construction will always yiels a polygroup.

Theorem 3.2.4. $\mathcal{A}[\mathcal{B}]$ *is a polygroup.*

Proof. The second condition of Definition 3.1.1 is clear. It is enough to check the conditions (P_1) and (P_3) of Definition 3.1.1. Without loss of generality we may assume $x, y, z \neq e$ and not all elements belong to A. Note that

(1) If $u \in B$ and $v \in A$, then $u * v = v * u = u$.

If exactly one of x, y, z belong to B, then (1) implies that both sides of (P_1) equal the element in $\{x, y, z\} \cap B$. If exactly two of x, y, z belong to B, say u and v, then (1) implies that both sides of (P_1) equal $u * v$. We assume that $x, y, z \in B \setminus \{e\}$ and show that

(2) $u \in (x * y) * z$ implies $u \in x * (y * z)$.

If $u \notin A$, then $u \in w * z$ for some $w \in x * y$. Now, if $w \notin A$, $w \in x \cdot y$ and $u \in w \cdot z$ so $u \in (xy)z = x(yz)$ (in B) $\subseteq x * (y * z)$. Also, if $w \in A$, $u \in w * z = z$ (so $u = z$) and $e \in xy$. Thus, $u = z \in (xy)z = x(yz) \subseteq x * (y * z)$. Now, suppose that $u \in A$. Then, $z^{-1} \in x * y$. Since $z^{-1} \notin A$ so $z^{-1} \in xy$ (in B), so $e \in (xy)z = x(yz)$. Thus, $x^{-1} \in y \cdot z \subseteq y * z$ and hence $u \in A \subseteq x * (y * z)$.

The proof of the opposite inclusion $x * (y * z) \subseteq (x * y) * z$. is similar to

(2).

The condition (P$_3$) is clear if $x, y, z \in A$. Since $x \in B \setminus \{e\}$ implies y or z belongs to $B \setminus \{e\}$ and $x \in A$ implies $z \in B \setminus \{e\}$, we may assume at least two of x, y, z belong to $B \setminus \{e\}$. On the other hand, if $x, y, z \in B \setminus \{e\}$, then $x \in y * z$ implies $x \in y \cdot z$ (in B) from which (P$_3$) follows. Therefore, we may assume exactly two of x, y, z belong to $B \setminus \{e\}$. This reduces to two cases:

(3) $x \in y * z$ where $x, y \in B \setminus \{e\}$ and $z \in A$.

By (1), $y * z = y$ so $x = y$. Thus, $y = x = x * z^{-1}$ using (1) again and $z \in A \subseteq x^{-1} * x = y^{-1} * x$.

(4) $x \in y * z$ where $x \in A$ and $y, z \in B \setminus \{e\}$.

In this case $y = z^{-1}$ so the desired conclusion follows using (1). This completes the proof of (P$_3$) and hence the theorem. ■

3.3 Subpolygroups and quotient polygroups

In this section, we study the concepts of subpolygroups and normal subpolygroups. In particular, by using the notion of the right coset of a subpolygroup, we show that there is a polygroup structure on the set of all right cosets of a given normal subpolygroup.

Definition 3.3.1. A non-empty subset K of a polygroup P is a *subpolygroup* of P if

(1) $a, b \in K$ implies $a \cdot b \subseteq K$,
(2) $a \in K$ implies $a^{-1} \in K$.

For simplicity of notations, sometimes we may write xy instead of $x \cdot y$.

Definition 3.3.2. A subpolygroup N of a polygroup P is *normal* in P if $a^{-1} N a \subseteq N$, for all $a \in P$.

The following corollaries are direct consequences of Definitions 3.3.1 and 3.3.2.

Corollary 3.3.3. *Let N be a normal subpolygroup of a polygroup P. Then,*

(1) $Na = aN$, for all $a \in P$;
(2) $(Na)(Nb) = Nab$, for all $a, b \in P$;
(3) $Na = Nb$, for all $b \in Na$.

Corollary 3.3.4. *Let K and N be subpolygroups of a polygroup P with N normal in P. Then,*

(1) *$N \cap K$ is a normal subpolygroup of K;*
(2) *$NK = KN$ is a subpolygroup of P;*
(3) *N is a normal subpolygroup of NK.*

Definition 3.3.5. If N is a normal subpolygroup of P, then we define the relation $x \equiv y(modN)$ if and only if $xy^{-1} \cap N \neq \emptyset$. This relation is denoted by $xN_P y$.

Lemma 3.3.6. *The relation N_P is an equivalence relation.*

Proof. (1) Since $e \in xx^{-1} \cap N$ for all $x \in P$; then $xN_P x$, i.e., N_P is reflexive. (2) Suppose that $xN_P y$. Then, there exists $z \in xy^{-1} \cap N$ which implies $z^{-1} \in yx^{-1}$ and $z^{-1} \in N$, this means that $yN_P x$, and so N_P is symmetric. (3) Let $xN_P y$ and $yN_P z$ where $x, y, z \in P$. Then, there exist $a \in xy^{-1} \cap N$ and $b \in yz^{-1} \cap N$. So $x \in ay$ and $z^{-1} \in y^{-1}b$, then $z^{-1}x \subseteq y^{-1}bay$. Since $ba \subseteq N$ and N is a normal subpolygroup, then $y^{-1}bay \subseteq N$. Therefore, $z^{-1}x \cap N \neq \emptyset$, which satisfies the condition for $xN_P z$, and so N_P is transitive. ∎

Let $N_P(x)$ be the equivalence class of the element $x \in P$. Suppose that $[P : N] = \{N_P(x) \mid x \in P\}$. On $[P : N]$ we consider the hyperoperation $\odot$ defined as follows: $N_P(x) \odot N_P(y) = \{N_P(z) \mid z \in N_P(x)N_P(y)\}$. For a subpolygroup K of P and $x \in P$, denote the right coset of K by Kx and let P/K is the set of all right cosets of K in P.

Lemma 3.3.7. *If N is a normal subpolygroup of P, then $Nx = N_P(x)$.*

Proof. Suppose that $y \in Nx$. Then, there exists $n \in N$ such that $y \in nx$, which implies that $n \in yx^{-1}$, and so $yx^{-1} \cap N \neq \emptyset$. Thus, $Nx \subseteq N_P(x)$. Similarly, we have $N_P(x) \subseteq Nx$. ∎

Therefore, we conclude that $[P : N] = P/N$.

Lemma 3.3.8. *Let N be a normal subpolygroup of P. Then, for all $x, y \in P$, $Nxy = Nz$ for all $z \in xy$.*

Proof. Suppose that $z \in xy$. Then, it is clear that $Nz \subseteq Nxy$. Now, let $a \in Nxy$. Then, we obtain $y \in (Nx)^{-1}a$ or $y \in x^{-1}Na$, and so $xy \subseteq xx^{-1}Na$. Since N is a normal subpolygroup, we obtain $xy \subseteq xNx^{-1}a \subseteq Na$. Therefore, for every $z \in xy$, we have $z \in Na$ which implies that $a \in Nz$. This completes the proof. ∎

Corollary 3.3.9. *For all* $x, y \in P$, *we have* $N_P(N_P(x)N_P(y)) = N_P(x)N_P(y)$.

Definition 3.3.10. An equivalence relation ρ on a polygroup P is called a *(full) conjugation* on P if

(1) $x\rho y$ implies $x^{-1}\rho y^{-1}$;
(2) $z \in xy$ and $z'\rho z$ implies $z' \in x'y'$ for some $x'\rho x$ and $y'\rho y$.

A conjugation ρ on P is a *special conjugation* if for all $x \in P$,

(3) $x\rho e$ implies $x = e$.

Using the notation of equivalence classes, a conjugation relation on P can be described, alternatively, as an equivalence relation ρ such that for all $x, y \in P$,

(1') $(\rho(x))^{-1} = \rho(x^{-1})$;
(2') $\rho(xy) \subseteq \rho(x)\rho(y)$.

A conjugation is special if $\rho(e) = \{e\}$.

Lemma 3.3.11. ρ *is a conjugation of* P *if and only if*

(1) $\rho(x)^{-1} = \{y^{-1} \mid y \in \rho(x)\} = \rho(x^{-1})$;
(2) $\rho(\rho(x)y) = \rho(x)\rho(y)$.

Proof. Assume that ρ is a conjugation. Condition (1) of Definition 3.3.10 easily implies (1) of lemma. Condition (2) of definition of conjugation implies that $\rho(x)\rho(y)$ is a union of ρ-classes so $\rho((\rho(x))y) \subseteq \rho(x)\rho(y)$. Now, assume that $z \in x' \cdot y'$, $x'\rho x$, and $y'\rho y$. By (P$_3$), $y' \in x'^{-1} \cdot z$ so $y \in x'' \cdot z'$, where $x'' \rho x'^{-1}$ and $z' \rho z$ by (2) of conjugation definition. By (P$_3$) again and (1),

$$z \rho z' \in (\rho(x)) \cdot y$$

which shows equality in (2). A similar argument establish that a ρ with properties (1) and (2) is a conjugation. ∎

Corollary 3.3.12. *The equivalence relation* N_P *is a conjugation on* P.

Proposition 3.3.13. $< [P : N], \odot, N_P(e), ^{-I} >$ *is a polygroup, where* $N_P(a)^{-I} = N_P(a^{-1})$.

Proof. For all $a, b, c \in P$, we have

$$
\begin{aligned}
(N_P(a) \odot N_P(b)) \odot N_P(c) &= \{N_P(x) \mid x \in N_P(a)N_P(b)\} \odot N_P(c) \\
&= \{N_P(y) \mid y \in N_P(x)N_P(c),\ x \in N_P(a)N_P(b)\} \\
&= \{N_P(y) \mid y \in N_P(N_P(a)N_P(b))N_P(c)\} \\
&= \{N_P(y) \mid y \in (N_P(a)N_P(b))N_P(c)\}, \\
N_P(a) \odot (N_P(b) \odot N_P(c)) &= N_P(a) \odot \{N_P(x) \mid x \in N_P(b)N_P(c)\} \\
&= \{N_P(y) \mid y \in N_P(a)N_P(x),\ x \in N_P(b)N_P(c)\} \\
&= \{N_P(y) \mid y \in N_P(a)N_P(N_P(b)N_P(c))\} \\
&= \{N_P(y) \mid y \in N_P(a)(N_P(b)N_P(c))\}.
\end{aligned}
$$

Since $(N_P(a)N_P(b))N_P(c) = N_P(a)(N_P(b)N_P(c))$, we get

$$
(N_P(a) \odot N_P(b)) \odot N_P(c) = N_P(a) \odot (N_P(b) \odot N_P(c)).
$$

Therefore, $\odot$ is associative. It is easy to see that $N_P(e)$ is the unit element in $[P : N]$, and $N_P(x^{-1})$ is the inverse of the element $N_P(x)$. Now, we show that $N_P(c) \in N_P(a) \odot N_P(b)$ implies $N_P(a) \in N_P(c) \odot N_P(b^{-1})$ and $N_P(b) \in N_P(a^{-1}) \odot N_P(c)$.

We have $N_P(c) \in N_P(a) \odot N_P(b)$, and hence $N_P(c) = N_P(x)$ for some $x \in N_P(a)N_P(b)$. Therefore, there exist $y \in N_P(a)$ and $z \in N_P(b)$ such that $x \in yz$, so $y \in xz^{-1}$. This implies that $N_P(y) \in N_P(x) \odot N_P(z^{-1})$, and so $N_P(a) \in N_P(c) \odot N_P(b^{-1})$. Similarly, we obtain $N_P(b) \in N_P(a^{-1}) \odot N_P(c)$. Therefore, $[P : N]$ is a polygroup. ∎

Corollary 3.3.14. *If N is a normal subpolygroup of P, then $< P/N, \odot, N, ^{-I} >$ is a polygroup, where $Nx \odot Ny = \{Nz \mid z \in xy\}$ and $(Nx)^{-I} = Nx^{-1}$.*

Definition 3.3.15. Let $< P_1, \cdot, e_1, ^{-1} >$ and $< P_2, *, e_2, ^{-I} >$ be polygroups. Let f be a mapping from P_1 into P_2 such that $f(e_1) = e_2$. Then, f is called

(1) an *inclusion homomorphism* if

$$
f(x \cdot y) \subseteq f(x) * f(y), \text{ for all } x, y \in P_1;
$$

(2) a *strong homomorphism* or a *good homomorphism* if

$$
f(x \cdot y) = f(x) * f(y), \text{ for all } x, y \in P_1;
$$

(3) a *homomorphism of type 2* if

$$
f^{-1}(f(x) * f(y)) = f^{-1}f(x \cdot y), \text{ for all } x, y \in P_1;
$$

(4) a *homomorphism of type 3* if

$$
f^{-1}(f(x) * f(y)) = (f^{-1}f(x)) \cdot (f^{-1}f(y)), \text{ for all } x, y \in P_1;
$$

(5) a *homomorphism of type 4* if

$$f^{-1}(f(x)*f(y)) = f^{-1}f(x \cdot y) = (f^{-1}f(x)) \cdot (f^{-1}f(y)), \text{ for all } x, y \in P_1;$$

where for non-empty subset A of P_1, by $f^{-1}f(A)$ we mean

$$f^{-1}f(A) = \bigcup_{x \in A} f^{-1}f(x).$$

Clearly, a strong homomorphism f is called an *isomorphism* if f is one to one and onto.

The defining condition for any types of homomorphism is also valid for sets. For instance, if f is an inclusion homomorphism of P_1 into P_2 and A, B are non-empty subsets of P_1, then it follows that $f(A \cdot B) \subseteq f(A) * f(B)$. Notice that a homomorphism of the types 2 through 4 is indeed an inclusion homomorphism. Observe that a one to one homomorphism of P_1 onto P_2 of any type 2 through 4 is an isomorphism.

The defining condition for a homomorphism of type 2 can easily be simplified if the homomorphism is onto.

Corollary 3.3.16. *Let f be a mapping of P_1 onto P_2. Then, f is a homomorphism of type 2 if and only if f is a strong homomorphism.*

The types of homomorphism introduced above have appeared in [86] and in the literature under various names. For example, homomorphisms of type 2 have been dealt with by Corsini [22], Koskas [94], Mittas [112], Prenowitz and Jantociak [123] and Sureau [138]. Also, Corsini [22] and Koskas [94] have also studied homomorphisms of type 3.

Why are factor polygroups important? Well, when P is finite and $N \neq \{e\}$, P/N is smaller than P, and its structure is usually less complicated than that of P. At the same time, P/N simulates P in many ways. In fact, we may think of a factor polygroup of P as a less complicated approximation of P. What makes factor polygroups important is that one can often deduce properties of P by examine the less complicated polygroup P/N instead. An excellent illustration of this is given in the next example.

Example 3.3.17. We illustrate constructions using the dihedral group D_4. This group is generated by a counter-clockwise rotation r of $90°$ and a horizontal reflection h. The group consists of the following 8 symmetries:

$$\{1 = r^0, r, r^2 = s, r^3 = t, h, hr = d, hr^2 = v, hr^3 = f\}.$$

The dihedral groups occur frequently in art and nature. Many of the decorative designs used on floor coverings, pottery, and buildings have one of the dihedral groups as a group of symmetry.

Let S_3 be the symmetric group of degree 3, i.e.,

$$S_3 = \{i, (1\ 2),\ (1\ 3),\ (2\ 3),\ (1\ 2\ 3),\ (1\ 3\ 2)\}$$

Let $\mathbb{A}$ denote the polygroup $S_3//(1\ 2)$. The multiplication table for $\mathbb{A}$ is

	A	B
A	A	B
B	B	$\{A, B\}$

Now, the polygroup $D_4 \times \mathbb{A}$ is isomorphic to $\wp$, where $\wp$ has the following multiplication table:

$\circ$	A_0	A_1	A_2	A_3	A_4	A_5	A_6	A_7	B_0	B_1	B_2	B_3	B_4	B_5	B_6	B_7
A_0	A_0	A_1	A_2	A_3	A_4	A_5	A_6	A_7	B_0	B_1	B_2	B_3	B_4	B_5	B_6	B_7
A_1	A_1	A_0	A_3	A_2	A_5	A_4	A_7	A_6	B_1	B_0	B_3	B_2	B_5	B_4	B_7	B_6
A_2	A_2	A_3	A_1	A_0	A_6	A_7	A_5	A_4	B_2	B_3	B_1	B_0	B_6	B_7	B_5	B_4
A_3	A_3	A_2	A_0	A_1	A_7	A_6	A_4	A_5	B_3	B_2	B_0	B_1	B_7	B_6	B_4	B_5
A_4	A_4	A_5	A_7	A_6	A_0	A_1	A_3	A_2	B_4	B_5	B_7	B_6	B_0	B_1	B_3	B_2
A_5	A_5	A_4	A_6	A_7	A_1	A_0	A_2	A_3	B_5	B_4	B_6	B_7	B_1	B_0	B_2	B_3
A_6	A_6	A_7	A_4	A_5	A_2	A_3	A_0	A_1	B_6	B_7	B_4	B_5	B_2	B_3	B_0	B_1
A_7	A_7	A_6	A_5	A_4	A_3	A_2	A_1	A_0	B_7	B_6	B_5	B_4	B_3	B_2	B_1	B_0
B_0	B_0	B_1	B_2	B_3	B_4	B_5	B_6	B_7	A_0, B_0	A_1, B_1	A_2, B_2	A_3, B_3	A_4, B_4	A_5, B_5	A_6, B_6	A_7, B_7
B_1	B_1	B_0	B_3	B_2	B_5	B_4	B_7	B_6	A_1, B_1	A_0, B_0	A_3, B_3	A_2, B_2	A_5, B_5	A_4, B_4	A_7, B_7	A_6, B_6
B_2	B_2	B_3	B_1	B_0	B_6	B_7	B_5	B_4	A_2, B_2	A_3, B_3	A_1, B_1	A_0, B_0	A_6, B_6	A_7, B_7	A_5, B_5	A_4, B_4
B_3	B_3	B_2	B_0	B_1	B_7	B_6	B_4	B_5	A_3, B_3	A_2, B_2	A_0, B_0	A_1, B_1	A_7, B_7	A_6, B_6	A_4, B_4	A_5, B_5
B_4	B_4	B_5	B_7	B_6	B_0	B_1	B_3	B_2	A_4, B_4	A_5, B_5	A_7, B_7	A_6, B_6	A_0, B_0	A_1, B_1	A_3, B_3	A_2, B_2
B_5	B_5	B_4	B_6	B_7	B_1	B_0	B_2	B_3	A_5, B_5	A_4, B_4	A_6, B_6	A_7, B_7	A_1, B_1	A_0, B_0	A_2, B_2	A_3, B_3
B_6	B_6	B_7	B_4	B_5	B_2	B_3	B_0	B_1	A_6, B_6	A_7, B_7	A_4, B_4	A_5, B_5	A_2, B_2	A_3, B_3	A_0, B_0	A_1, B_1
B_7	B_7	B_6	B_5	B_4	B_3	B_2	B_1	B_0	A_7, B_7	A_6, B_6	A_5, B_5	A_4, B_4	A_3, B_3	A_2, B_2	A_1, B_1	A_0, B_0

Let $N = \{A_0, A_1, B_0, B_1\}$, and consider the factor polygroup $\wp$ by N,

$$\wp/N = \{N, A_2N, A_4N, A_6N\}$$

The multiplication table for $\wp/N$ is given in the following table:

	N	A_2N	A_4N	A_6N
N	N	A_2N	A_4N	A_6N
A_2N	A_2N	N	A_6N	A_4N
A_4N	A_4N	A_6N	N	A_2N
A_6N	A_6N	A_4N	A_2N	N

$\wp/N$ provides a good opportunity to demonstrate how a factor polygroup of P is related to P itself. Suppose that we arrange the heading of the Cayley table for $\wp$ in such a way that elements from the same coset of N are in adjacent columns. Then, the multiplication table for $\wp$ can be blocked off into boxes which are cosets of N, and the substitution that replaces a box containing the element X with the coset XN yields the Cayley table for $\wp/N$.

$\circ$	A_0	A_1	B_0	B_1	A_2	A_3	B_2	B_3	A_4	A_5	B_4	B_5	A_6	A_7	B_6	B_7
A_0	A_0	A_1	B_0	B_1	A_2	A_3	B_2	B_3	A_4	A_5	B_4	B_5	A_6	A_7	B_6	B_7
A_1	A_1	A_0	B_1	B_0	A_3	A_2	B_3	B_2	A_5	A_4	B_5	B_4	A_7	A_6	B_7	B_6
B_0	B_0	B_1	A_0,B_0	A_1,B_1	B_2	B_3	A_2,B_2	A_3,B_3	B_4	B_5	A_4,B_4	A_5,B_5	B_6	B_7	A_6,B_6	A_7,B_7
B_1	B_1	B_0	A_1,B_1	A_0,B_0	B_3	B_2	A_3,B_3	A_2,B_2	B_5	B_4	A_5,B_5	A_4,B_4	B_7	B_6	A_7,B_7	A_6,B_6
A_2	A_2	A_3	B_2	B_3	A_1	A_0	B_1	B_0	A_6	A_7	B_6	B_7	A_5	A_4	B_5	B_4
A_3	A_3	A_2	B_3	B_2	A_0	A_1	B_0	B_1	A_7	A_6	B_7	B_6	A_4	A_5	B_4	B_5
B_2	B_2	B_3	A_2,B_2	A_3,B_3	B_1	B_0	A_1,B_1	A_0,B_0	B_6	B_7	A_6,B_6	A_7,B_7	B_5	B_4	A_5,B_5	A_4,B_4
B_3	B_3	B_2	A_3,B_3	A_2,B_2	B_0	B_1	A_0,B_0	A_1,B_1	B_7	B_6	A_7,B_7	A_6,B_6	B_4	B_5	A_4,B_4	A_5,B_5
A_4	A_4	A_5	B_4	B_5	A_7	A_6	B_7	B_6	A_0	A_1	B_0	B_1	A_3	A_2	B_3	B_2
A_5	A_5	A_4	B_5	B_4	A_6	A_7	B_6	B_7	A_1	A_0	B_1	B_0	A_2	A_3	B_2	B_3
B_4	B_4	B_5	A_4,B_4	A_5,B_5	B_7	B_6	A_7,B_7	A_6,B_6	B_0	B_1	A_0,B_0	A_1,B_1	B_3	B_2	A_3,B_3	A_2,B_2
B_5	B_5	B_4	A_5,B_5	A_4,B_4	B_6	B_7	A_6,B_6	A_7,B_7	B_1	B_0	A_1,B_1	A_0,B_0	B_2	B_3	A_2,B_2	A_3,B_3
A_6	A_6	A_7	B_6	B_7	A_4	A_5	B_4	B_5	A_2	A_3	B_2	B_3	A_0	A_1	B_0	B_1
A_7	A_7	A_6	B_7	B_6	A_5	A_4	B_5	B_4	A_3	A_2	B_3	B_2	A_1	A_0	B_1	B_0
B_6	B_6	B_7	A_6,B_6	A_7,B_7	B_4	B_5	A_4,B_4	A_5,B_5	B_2	B_3	A_2,B_2	A_3,B_3	B_0	B_1	A_0,B_0	A_1,B_1
B_7	B_7	B_6	A_7,B_7	A_6,B_6	B_5	B_4	A_5,B_5	A_4,B_4	B_3	B_2	A_3,B_3	A_2,B_2	B_1	B_0	A_1,B_1	A_0,B_0

Thus, when we pass from $\wp$ to $\wp/N$, the box

A_0	A_1	B_0	B_1
A_1	A_0	B_1	B_0
B_0	B_1	A_0,B_0	A_1,B_1
B_1	B_0	A_1,B_1	A_0,B_0

in Table 3.3.3 becomes the element N in Table 3.3.2, similarly, the box

A_3	A_2	B_3	B_2
A_2	A_3	B_2	B_3
B_3	B_2	A_3,B_3	A_2,B_2
B_2	B_3	A_2,B_2	A_3,B_3

becomes the element A_2N, and so on.

In this way, one can see that the formation of a factor polygroup $\wp/N$ causes a systematic collapsing of the elements X collapse to the single polygroup element XN in $\wp/N$.

3.4   Isomorphism theorems of polygroups

We recall that a strong homomorphism $\varphi : P_1 \longrightarrow P_2$ is an *isomorphism* if φ is one to one and onto. We write $P_1 \cong P_2$ if P_1 is isomorphic to P_2.

Because P_1 is a polygroup, $e \in aa^{-1}$ for all $a \in P_1$, then we have $\varphi(e_1) \in \varphi(a) * \varphi(a^{-1})$ or $e_2 \in \varphi(a) * \varphi(a^{-1})$ which implies $\varphi(a^{-1}) \in \varphi(a)^{-1} * e_2$. Therefore, $\varphi(a^{-1}) = \varphi(a)^{-1}$ for all $a \in P_1$. Moreover, if φ is a strong homomorhism from P_1 into P_2, then the kernel of φ is the set $ker\varphi = \{x \in P_1 \mid \varphi(x) = e_2\}$. It is trivial that $ker\varphi$ is a subpolygroup of P_1 but in general is not normal in P_1.

Lemma 3.4.1. *Let φ be a strong homomorphism from P_1 into P_2. Then, φ is injective if and only if $ker\varphi = \{e_1\}$.*

Proof. Let $y, z \in P_1$ be such that $\varphi(y) = \varphi(z)$ Then, $\varphi(y) * \varphi(y^{-1}) = \varphi(z) * \varphi(y^{-1})$. It follows that $\varphi(e_1) \in \varphi(yy^{-1}) = \varphi(zy^{-1})$, and so there exists $x \in yz^{-1}$ such that $e_2 = \varphi(e_1) = \varphi(x)$. Thus, if $ker\varphi = \{e_1\}$, $x = e_1$, whence $y = z$. Now, let $x \in ker\varphi$. Then, $\varphi(x) = e_2 = \varphi(e_1)$. Thus, if φ is injective, we conclude that $x = e_1$. ∎

We are now in a position to state and review the fundamental theorems in polygroup theory.

Theorem 3.4.2 (First Isomorphism Theorem). *Let φ be a strong homomorphism from P_1 into P_2 with kernel K such that K is a normal subpolygroup of P_1. Then, $P_1/K \cong Im\varphi$.*

Proof. We define $\psi : P_1/K \longrightarrow Im\varphi$ by setting $\psi(Kx) = \varphi(x)$ for all $x \in P_1$. It is easy to see that ψ is an isomorphism. ∎

Theorem 3.4.3. (Second Isomorphism Theorem). *If K and N are subpolygroups of a polygroup P, with N normal in P, then $K/(N \cap K) \cong NK/N$.*

Proof. Since N is a normal subpolygroup of P, $NK = KN$. Consequently NK is a subpolygroup of P. Further $N = Ne \subseteq NK$ given that N is a normal subpolygroup of NK; consequently NK/N is defined. Define $\varphi : K \longrightarrow NK/N$ by $\varphi(k) = Nk$. φ is a strong homomorphism. Consider any $Na \in NK/N$, $a \in NK$. Now $a \in NK$ given $a \in nk$ for some $n \in N$, $k \in K$. Thus, by Corollary 3.3.3, $Na = Nnk = Nk = \varphi(k)$. This shows that φ is also onto. If we can establish that $ker\varphi = N \cap K$, since $N \cap K$ is a normal subpolygroup of K, we shall get that $K/N \cap K \cong NK/N$. For any $k \in K$, $k \in ker\varphi \iff \varphi(k) = N \iff Nk = N \iff k \in N \iff k \in N \cap K$ (since $k \in K$), i.e., $k \in ker\varphi \iff k \in N \cap K$. This yields $ker\varphi = N \cap K$. Hence, that results follows. ∎

Theorem 3.4.4. (Third Isomorphism Theorem). *If K and N are normal subpolygroups of a polygroup P such that $N \subseteq K$, then K/N is a normal subpolygroup of P/N and $(P/N)/(K/N) \cong P/K$.*

Proof. We leave it to reader to verify that K/N is a normal subpolygroup of P/N. Further $\varphi : P/N \longrightarrow P/K$ defined by $\varphi(Nx) = Kx$ is a strong homomorphism of P/N onto P/K such that $ker\varphi = K/N$. ∎

Corollary 3.4.5. *If N_1, N_2 are normal subpolygroups of P_1, P_2 respectively, then $N_1 \times N_2$ is a normal subpolygroup of $P_1 \times P_2$ and $(P_1 \times P_2)/(N_1 \times N_2) \cong P_1/N_1 \times P_2/N_2$.*

Let P be a polygroup. We define the relation β^* as the smallest equivalence relation on P such that the quotient P/β^*, the set of all equivalence classes, is a group. In this case β^* is called the *fundamental equivalence relation* on P and P/β^* is called the *fundamental group*. The product $\otimes$ in P/β^* is defined as follows: $\beta^*(x) \otimes \beta^*(y) = \beta^*(z)$ for all $z \in \beta^*(x)\beta^*(y)$. Let $\mathcal{U}_P$ be the set of all finite products of elements of P. We define the relation β as follows: $x \, \beta \, y$ if and only if $\{x, y\} \subseteq u$ for some $u \in \mathcal{U}_P$. We have $\beta^* = \beta$ for hypergroups. Since polygroups are certain subclasses of hypergroups, we have $\beta^* = \beta$. The kernel of the canonical map $\varphi : P \longrightarrow P/\beta^*$ is called the *core* of P and is denoted by ω_P. Here we also denote by ω_P the unit of P/β^*. It is easy to prove that the following statements: $\omega_P = \beta^*(e)$ and $\beta^*(x)^{-1} = \beta^*(x^{-1})$ for all $x \in P$.

Theorem 3.4.6. *Let β_1^*, β_2^* and β^* be fundamental equivalence relations on polygroups P_1, P_2 and $P_1 \times P_2$ respectively, then $(P_1 \times P_2)/\beta^* \cong P_1/\beta_1^* \times P_2/\beta_2^*$.*

Corollary 3.4.7. *If N_1, N_2 are normal subpolygroups of P_1, P_2 respectively, and β_1^*, β_2^* and β^* fundamental equivalence relations on P_1/N_1, P_2/N_2 and $(P_1 \times P_2)/(N_1 \times N_2)$ respectively, then*

$$((P_1 \times P_2)/(N_1 \times N_2))/\beta^* \cong (P_1/N_1)/\beta_1^* \times (P_2/N_2)/\beta_2^*.$$

Definition 3.4.8. Let f be a strong homomorphism from P_1 into P_2 and let β_1^*, β_2^* be fundamental relations on P_1, P_2 respectively. We define

$$\overline{\ker f} = \{\beta_1^*(x) \mid x \in P_1, \ \beta_2^*(f(x)) = \omega_{P_2}\}.$$

Lemma 3.4.9. $\overline{\ker f}$ *is a normal subgroup of the fundamental group P_1/β_1^*.*

Proof. If $\beta_1^*(x), \beta_1^*(y) \in \overline{\ker f}$, then for every $z \in xy^{-1}$ we have $\beta_1^*(z) = \beta_1^*(x) \otimes \beta_1^*(y^{-1})$. On the other hand, we have

$$\beta_2^*(f(z)) = \beta_2^*(f(x)f(y^{-1})) = \beta_2^*(f(x)) \otimes \beta_2^*(f(y^{-1})) = \omega_{P_2} \otimes \omega_{P_2} = \omega_{P_2}.$$

Therefore, $\beta_1^*(z) \in \overline{\ker f}$. Now, let $\beta_1^*(a) \in P_1/\beta_1^*$ and $\beta_1^*(x) \in \overline{\ker f}$. Then, for every $z \in axa^{-1}$ we have $\beta_1^*(z) = \beta_1^*(a) \otimes \beta_1^*(x) \otimes \beta_1^*(a^{-1})$. On the other hand, we have

$$
\begin{aligned}
\beta_2^*(f(z)) &= \beta_2^*(f(a)f(x)f(a^{-1})) \\
&= \beta_2^*(f(a)) \otimes \beta_2^*(f(x)) \otimes \beta_2^*(f(a^{-1})) \\
&= \beta_2^*(f(a)) \otimes \omega_{P_2} \otimes \beta_2^*(f(a^{-1})) \\
&= \beta_2^*(f(aa^{-1})) = \beta_2^*(f(e_1)) = \beta_2^*(e_2) = \omega_{P_2}.
\end{aligned}
$$

Hence, we obtain $\beta_1^*(z) \in \overline{\ker f}$. This completes the proof. ∎

Theorem 3.4.10. *Let P be a polygroup, M, N be two normal subpolygroups of P with $N \subseteq M$ and $\phi : P/N \longrightarrow P/M$ be the canonical map. Suppose that β_M^*, β_N^* are the fundamental equivalence relations on P/M, P/N, respectively. Then, $((P/N)/\beta_N^*)/\overline{ker\phi} \cong (P/M)/\beta_M^*$.*

Proof. We define the map $\psi : (P/N)/\beta_N^* \longrightarrow (P/M)/\beta_M^*$ by $\psi : \beta_N^*(Nx) \longmapsto \beta_M^*(Mx)$ (for all $x \in P$). We must check that ψ is well-defined, that is, if $x, y \in P$ and $\beta_N^*(Nx) = \beta_N^*(Ny)$ then $\beta_M^*(Mx) = \beta_M^*(My)$. Now $\beta_N^*(Nx) = \beta_N^*(Ny)$ if and only if $\{Nx, Ny\} \subseteq u$ for some $u \in \mathcal{U}_{P/N}$. We have $u = Nx_1 \odot Nx_2 \odot \ldots \odot Nx_n = \{Nz \mid z \in \prod_{i=1}^{n} x_i\}$. Therefore, for some $z_1 \in \prod_{i=1}^{n} x_i$, $z_2 \in \prod_{i=1}^{n} x_i$ we have $Nx = Nz_1$ and $Ny = Nz_2$. So there exist $a \in xz_1^{-1} \cap N$ and $b \in yz_2^{-1} \cap N$, then $x \in az_1$ and $y \in bz_2$. Hence, $Mx \in Ma \odot Mz_1$ and $My \in Mb \odot Mz_2$. Since $a, b \in N \subseteq M$, then $Ma = M$, $Mb = M$. Since $M \odot Mz_1 = Mz_1$ and $M \odot Mz_2 = Mz_2$, we have $Mx = Mz_1$ and $My = Mz_2$. From $\{Mz_1, Mz_2\} \subseteq \{Mz \mid z \in \prod_{i=1}^{n} x_i\}$, we get $\{Mx, My\} \subseteq \{Mz \mid z \in \prod_{i=1}^{n} x_i\} = Mx_1 \odot Mx_2 \odot \ldots \odot Mx_n$. Therefore, $\beta_M^*(Mx) = \beta_M^*(My)$. This follows that ψ is well-defined. Moreover, ψ is a strong homomorphism, for if $x, y \in P$. Then,

$$\psi(\beta_N^*(Nx) \otimes \beta_N^*(Ny)) = \psi(\beta_N^*(Nxy)) = \beta_M^*(Mxy)$$
$$= \beta_M^*(Mx) \otimes \beta_M^*(My)$$
$$= \psi(\beta_N^*(Nx)) \otimes \psi(\beta_M^*(My)),$$

and $\psi(\omega_{P/N}) = \psi(\beta_N^*(N)) = \beta_M^*(M) = \omega_{P/M}$. Clearly, ψ is onto. Now, we show that $ker\psi = \overline{ker\phi}$. We have

$$ker\psi = \{\beta_N^*(Nx) \mid \psi(\beta_N^*(Nx)) = \omega_{P/N}\}$$
$$= \{\beta_N^*(Nx) \mid \beta_M^*(Mx) = \omega_{P/N}\}$$
$$= \{\beta_N^*(Nx) \mid \beta_M^*(\phi(Nx)) = \omega_{P/N}\}$$
$$= \overline{ker\phi}. \blacksquare$$

3.5 γ^* relation on polygroups

We recall the following definition. If H is a semihypergroup, then we set: $\gamma_1 = \{(x, x) \mid x \in H\}$ and, for every integer $n > 1$, the relation is defined as follows:

$$x\gamma_n y \iff \exists(z_1, z_2, \ldots, z_n) \in H^n, \exists \sigma \in \mathbb{S}_n : x \in \prod_{i=1}^{n} z_i, \ y \in \prod_{i=1}^{n} z_{\sigma(i)}.$$

Obviously, for every $n \geq 1$, the relations γ_n are symmetric, and the relation $\gamma = \bigcup_{n \geq 1} \gamma_n$ is reflexive and symmetric. Let γ^* be the *transitive closure* of γ. Then, γ^* is the smallest strongly regular equivalence such that H/γ^* is a commutative semigroup. Also, in every hypergroup, the relation γ is transitive, that is $\gamma^* = \gamma$, and in this case, the quotient H/γ^* is a commutative group. The γ^*-relation is a generalization of the β^*-relation.

Let P be a polygroup. We consider the relation γ^* on P. The product $\otimes$ in P/γ^* is defined as follows: $\gamma^*(x) \otimes \gamma^*(y) = \gamma^*(z)$ for all $z \in \gamma^*(x)\gamma^*(y)$. Clearly, we have $\gamma^*(e) = 1_{P/\gamma^*}$ and $(\gamma^*(x))^{-1} = \gamma^*(x^{-1})$ for all $x \in P$. Let $\phi : P \longrightarrow P/\gamma^*$ be the canonical projection. D_P is called the *derived hypergroup* and we have $D_P = \phi^{-1}(1_{P/\gamma^*})$. For every non-empty subset M of polygroup P we have $\phi^{-1}(\phi(M)) = D_P M = M D_P$.

Theorem 3.5.1. *Let Γ^* and $\gamma_{\mathcal{B}}^*$ be the γ^*-relations defined on $\mathcal{A}[\mathcal{B}]$ and $\mathcal{B}$. Then, we have*

$$\mathcal{A}[\mathcal{B}]/\Gamma^* \cong \mathcal{B}/\gamma_{\mathcal{B}}^*.$$

Proof. Note that all the elements of A belongs to the class $\Gamma^*(e)$. ∎

Example 3.5.2. Let S_3 be the symmetric group of degree 3 and let $\mathcal{A}$ denote the polygroup $S_3//(1\ 2)$ (*double coset algebra*). The multiplication table for $\mathcal{A}$ is

	A	B
A	A	B
B	B	$\{A, B\}$

Now, let the polygroup $\mathcal{B}$ has the following multiplication table:

	e	a	b	c	d
e	e	a	b	c	d
a	a	e	b	c	d
b	b	b	$\{e, a\}$	d	c
c	c	c	d	$\{e, a\}$	b
d	d	d	c	b	$\{e, a\}$

Then, $\mathcal{A}[\mathcal{B}]$ is a polygroup. By Theorem 3.5.1, the Γ^*-classes are $x = \{e, a\}$, $y = \{b\}$, $z = \{c\}$ and $t = \{d\}$. Therefore, $\mathcal{A}[\mathcal{B}]/\Gamma^*$ is the Klein's group.

Theorem 3.5.3. *Let P be a polygroup and $a_1, \ldots, a_k, b_1, \ldots, b_k \in P$ such that $a_j \gamma^* b_j$ for all $j = 1, \ldots k$. Then, for all $x \in a_1^{\delta_1} \ldots a_k^{\delta_k}$ and for all $y \in b_1^{\delta_1} \ldots b_k^{\delta_k}$ where $\delta_i \in \{1, -1\}$ $(i = \{1, \ldots, k\})$, we have $x \gamma^* y$.*

Proof. Suppose that $a_j \gamma^* b_j$ for all $j = 1, \ldots, k$. Then, there exist $n_j \in \mathbb{N}$ and $(z_{j1}, \ldots, z_{jn_j}) \in P^{n_j}$, and there exists $\sigma_j \in \mathbb{S}_{n_j}$ such that

$$a_j \in \prod_{i=1}^{n_j} z_{ji} \text{ and } b_j \in \prod_{i=1}^{n_j} z_{j\sigma_j(i)}.$$

Therefore,

$$a_1 \ldots a_k \subseteq \prod_{i=1}^{n_1} z_{1i} \cdots \prod_{i=1}^{n_k} z_{ki} \text{ and } b_1 \ldots b_k \subseteq \prod_{i=1}^{n_1} z_{1\sigma_1(i)} \cdots \prod_{i=1}^{n_k} z_{k\sigma_k(i)}.$$

If we rename z_{ij}'s, then we conclude that there exists $\Sigma \in \mathbb{S}_{n_1 + \ldots + n_k}$ such that

$$a_1 \ldots a_k \subseteq \prod_{i=1}^{n_1 + \ldots + n_k} z_i \text{ and } b_1 \ldots b_k \subseteq \prod_{i=1}^{n_1 + \ldots + n_k} z_{\Sigma(i)},$$

and so for all $x \in a_1 \ldots a_k$ and for all $y \in b_1 \ldots b_k$ we obtain $x \gamma^* y$. Now, note that if $a_j \gamma^* b_j$, then $a_j^{-1} \gamma^* b_j^{-1}$. ∎

The relational notation $A \approx B$ is used to assert that the sets A and B have at least one element in common.

Theorem 3.5.4. *Let P be a polygroup. Then, $x \gamma^* y$ if and only if there exist $A, A' \subseteq \gamma^*(a)$ and $B, B' \subseteq \gamma^*(b)$ for some $a, b \in P$ such that $xA \approx B$ and $yA' \approx B'$.*

Proof. Suppose that there exist $A, A' \subseteq \gamma^*(a)$ and $B, B' \subseteq \gamma^*(b)$ for some $a, b \in P$ such that $xA \approx B$ and $yA' \approx B'$. Then, we have

$$(\gamma^*(x) \otimes \{\gamma^*(y') \mid y' \in A\}) \approx \{\gamma^*(z) \mid z \in B\},$$

$$(\gamma^*(y) \otimes \{\gamma^*(y'') \mid y'' \in A'\}) \approx \{\gamma^*(z'') \mid z'' \in B'\}.$$

Therefore, we obtain $\gamma^*(x) \otimes \gamma^*(a) = \gamma^*(b)$ and $\gamma^*(y) \otimes \gamma^*(a) = \gamma^*(b)$, which implies that $\gamma^*(x) = \gamma^*(y) = \gamma^*(b) \otimes \gamma^*(a^{-1})$. Therefore, $x \gamma^* y$.

For the converse, if $x \gamma^* y$, then we can take $A = A' = D_P$ and $B = B' = \gamma^*(x)$. This completes the proof. ∎

Corollary 3.5.5. *Let P be a polygroup. Then, $x, y \in \gamma^*(e)$ if and only if there exist $A, A' \subseteq \gamma^*(z)$ and $B, B' \subseteq \gamma^*(z)$ for some $z \in P$ such that $xA \approx B$ and $yA' \approx B'$.*

Theorem 3.5.6. *Let P be a finite polygroup. For every $a \in P$, there exists a power a^r, we take the minimal one, which contains an element of a lower power, that is, there exists a^s such that $a^r \approx a^s$, $0 < s < r$. Then, $a^{r-s} \subseteq \gamma^*(e)$.*

Proof. From $a^r \approx a^s$ we have $\phi(a^r) = \phi(a^s)$, and so $(\gamma^*(a))^r = (\gamma^*(a))^s$. Since $(\gamma^*(a))^r$ and $(\gamma^*(a))^s$ are the elements of abelian group P/γ^*, then

$$(\gamma^*(a))^{r-s} = 1_{P/\gamma^*} = \gamma^*(e)$$

which implies that $\phi(a^{r-s}) = \gamma^*(e)$, and so $a^{r-s} \subseteq \gamma^*(e)$. ∎

Let M be a non-empty subset of a polygroup P, we say that M is a γ-*part* of P, if for every $n \in \mathbb{N}$, for every $(z_1, \ldots, z_n) \in P^n$ and for every $\sigma \in \mathbb{S}_n$, we have:

$$\prod_{i=1}^n z_i \approx M \Longrightarrow \prod_{i=1}^n z_{\sigma(i)} \subseteq M.$$

Let A be a non-empty subset of P. The intersection of γ-parts P which contain A is called the γ-*closure* of A in P. It will be denoted $C_\gamma(A)$. We have

    (1) $A \in \mathcal{P}^*(P)$, one has $D_P A = A D_P = C_\gamma(A)$.
    (2) $A \in \mathcal{P}^*(P)$, then A is a γ-part of P if and only if $A D_P = A$.
    (3) If A is a γ-part of P then AB, BA are γ-parts of P for every $B \in \mathcal{P}^*(P)$.
    (4) D_P is a γ-part of P.

Let P be a polygroup and $\prod(P)$ be the set of hyperproducts of elements of P. Let $X =< \prod(P), \odot >$ be the set of non-empty subsets of P endowed with the hyperoperation $\odot$ defined as follows:

$$A \odot B = \{ C \in \prod(P) \mid C \subseteq AB \},$$

for all $A, B \in \prod(P)$ with $A, B \neq \{e\}$, and $A \odot \{e\} = \{e\} \odot A = A$. Then, we have

Theorem 3.5.7. *If P is a polygroup, then $< \prod(P), \odot >$ is a regular hypergroup.*

Proof. First, we show that $\odot$ on $\prod(P)$ is associative. Let $A_1, A_2, A_3 \in \prod(P)$. Then, we have

$$\begin{aligned}
A_1 \odot (A_2 \odot A_3) &= A_1 \odot \{ C \in \prod(P) \mid C \subseteq A_2 A_3 \} \\
&= \{ D \in \prod(P) \mid D \subseteq A_1(A_2 A_3) \} \\
&= \{ D \in \prod(P) \mid D \subseteq (A_1 A_2) A_3 \} \\
&= \{ C \in \prod(P) \mid C \subseteq A_1 A_2 \} \odot A_3 \\
&= (A_1 \odot A_2) \odot A_3.
\end{aligned}$$

Let's prove now the reproducibility. Let $A = \prod_{i=1}^p a_i$ and $B = \prod_{i=1}^q b_i$ be elements of $\prod(P)$. By reproducibility of $\cdot$, there exists $y_1 \in P$ such that

$a_p \in y_1 \cdot b_q$. Similarly, there is y_2 such that $y_1 \in y_2 \cdot b_{q-1}$, whence $a_p \in y_2 \cdot b_{q-1} \cdot b_q$. Going up in the same way, one obtains y_q such that $y_{q-1} \in y_q \cdot b_1$. Hence, $a_p \in y_q \cdot b_1 \cdot b_2 \cdot \ldots \cdot b_q$. Therefore, if we let $X = \prod_{i=1}^{p-1} a_i \cdot y_q$, we have $A \in X \odot B$.

Similarly, we can find $z_1, z_2, \ldots, z_q$ such that $a_1 \in b_1 \cdot z_1$, $z_1 \in b_2 \cdot z_2, \ldots, z_{q-1} \in b_q \cdot z_q$, whence $A = a_1 \cdot a_2 \cdot \ldots \cdot a_p \subseteq b_1 \cdot b_2 \cdot \ldots \cdot b_q \cdot z_q \cdot a_1 \cdot a_2 \cdot \ldots \cdot a_p$.

Now, let $E = \{e\}$, then for all $A \in \prod(P)$ we have $A \odot E = E \odot A = A$. We define the unary operation $^{-I}$ as follows:

$$^{-I} : \prod(P) \longrightarrow \prod(P)$$
$$(x_1 \ldots x_n)^{-I} = x_n^{-1} \ldots x_1^{-1}. \quad \blacksquare$$

Corollary 3.5.8. *If A is a subpolygroup of P and A belongs to $\prod(P)$, then A is contained in D_P.*

The following example show that not all subpolygroups of a polygroup P are in $\prod(P)$.

Example 3.5.9. Let P be a polygroup with the following table:

	a	b	c	d
a	a	b	c	d
b	b	a	c	d
c	c	c	$\{a,b,d\}$	$\{c,d\}$
d	d	d	$\{c,d\}$	$\{a,b,c\}$

It is clear that $A = \{a, b\}$ is a subpolygroup of P, but $A \notin \prod(P)$. Moreover $D_P = cdd = P \in \prod(P)$.

Note that the following example shows that, in general, by Theorem 3.5.7, we do not obtain a polygroup structure.

Example 3.5.10. Let $P = \{a, b\}$ with $a^{-1} = a$, $b^{-1} = b$ and the following hyperoperation

	a	b
a	a	b
b	b	$\{a,b\}$

If $X = a \cdot a$, $Y = b \cdot b$ and $Z = a \cdot a$, then $X \in Y \odot Z$ but $Y \notin X \odot z^{-I}$.

Indeed, Theorem 6 in [37] holds, if we add the following condition in the definition of $\odot$:

(∗) $X \in A \odot B$, if for every $a \in A$ and $b \in B$, there exists $x \in X$ such that $x \in a \cdot b$.

Theorem 3.5.11. *Let P be a polygroup. Then $< \prod(P), \odot >$ is a poly-group, if $\odot$ satisfies in the condition $(*)$.*

Proof. Now, let $X = \prod\limits_{i=1}^{m} x_i$, $Y = \prod\limits_{i=1}^{n} y_i$ and $Z = \prod\limits_{i=1}^{p} z_i$ be elements of $\prod(P)$ such that $X \in Y \odot Z$. Hence, $X \subseteq Y \cdot Z$ and for every $y \in Y$ and $z \in Z$ there exists $x \in X$ such that $x \in y \cdot z$ and this implies that $y \in xz^{-1}$, and so $y \in (x_1 \ldots x_n)(z_p^{-1} \ldots z_1^{-1})$ for every $y \in Y$. Therefore, $Y \subseteq (x_1 \ldots x_n)(z_p^{-1} \ldots z_1^{-1})$ or $Y \in X \odot Z^{-I}$. Similarly, we get $Z \in Y^{-I} \odot X$. ∎

If P is a polygroup, we denote $\prod\limits_{C_\gamma}(P)$ the set of hyperproducts of elements of P, such that $C_\gamma(A) = A$, for all $A \in \prod\limits_{C_\gamma}(P)$.

Theorem 3.5.12. *Let P be a polygroup and $(x_1, \ldots, x_n) \in P^n$ be such that $\prod\limits_{i=1}^{n} x_i \in \prod\limits_{C_\gamma}(P)$. Then, there exists $(y_1, \ldots, y_n) \in P^n$ such that $x_1 \ldots x_n y_n \ldots y_1 = D_P$.*

Proof. For $1 \leq j \leq n$, let a_j be an element of D_P. Then, there exists $y_j \in P$ such that $a_j \in x_j y_j$. Since D_P is a γ-part, we have $x_j y_j \subseteq D_P$. Therefore,

$$
\begin{aligned}
x_1 \ldots x_n y_n &= D_P x_1 \ldots x_n y_n \\
&= x_1 \ldots x_{n-1} D_P x_n y_n \\
&= x_1 \ldots x_{n-1} D_P \\
&= D_P \prod_{i=1}^{n-1} x_i,
\end{aligned}
$$

and so

$$
\begin{aligned}
x_1 \ldots x_n y_n y_{n-1} &= D_P \left(\prod_{i=1}^{n-2} x_i \right) x_{n-1} y_{n-1} \\
&= D_P \prod_{i=1}^{n-2} x_i.
\end{aligned}
$$

Going on in the same way, one arrives to $x_1 \ldots x_n y_n \ldots y_2 = D_P x_1$ whence finally

$$
x_1 \ldots x_n y_n \ldots y_1 = D_P x_1 y_1 = D_P. \quad ∎
$$

Theorem 3.5.13. *Let P be a polygroup. If $P \backslash D_P$ is a hyperproduct, then D_P is a hyperproduct, too.*

Proof. Since D_P is a γ-part, so $P \backslash D_P$ is also a γ-part. Now, by using

Theorem 3.5.12, the proof is completed. ∎

Using the γ^*-relation we define the semidirect hyperproducts of polygroups.

Definition 3.5.14. Let $\mathbb{A} =< A, \cdot, e_1, ^{-1} >$ and $\mathbb{B} =< B, *, e_2, ^{-1} >$ be two polygroups. We consider the group $Aut A$ and the group B/γ_B^*, let

$$\widehat{}: B/\gamma_B^* \longrightarrow Aut A$$

$$\gamma_B^*(b) \longrightarrow \widehat{\gamma_B^*(b)} = \widehat{b}$$

be a homomorphism of groups. Then, in $\mathbb{A} \times \mathbb{B}$ we define a hyperproduct as follows:

$$(a_1, b_1) \circ (a_2, b_2) = \{(x, y) | x \in a_1 \cdot \widehat{b_1}(a_2), y \in b_1 * b_2\}$$

and we call this the *semidirect hyperproduct* of polygroups $\mathbb{A}$ and $\mathbb{B}$.

The above definition, first was introduced by Vougiouklis concerning hypergroups and the fundamental relation β^*.

Theorem 3.5.15. $\mathbb{A} \times \mathbb{B}$ *equipped with the semidirect hyperproduct is a polygroup.*

Proof. It is easy to see that the associativity is valid. Since $a = a \cdot \widehat{b}(e_1)$, $a = e_1 \cdot \widehat{e_2}(a)$ and $b = b * e_2 = e_2 * b$, we have $(a, b) \circ (e_1, e_2) = (a, b) = (e_1, e_2) \circ (a, b)$, that is, (e_1, e_2) is the identity element in $\mathbb{A} \times \mathbb{B}$, and we can check that $(\widehat{b^{-1}}(a^{-1}), b^{-1})$ is the inverse element of (a, b) in $\mathbb{A} \times \mathbb{B}$. Now, we show that $(z_1, z_2) \in (x_1, x_2) \circ (y_1, y_2)$ imply $(x_1, x_2) \in (z_1, z_2) \circ (y_1, y_2)^{-I}$ and $(y_1, y_2) \in (x_1, x_2)^{-I} \circ (z_1, z_2)$. We have

$$(z_1, z_2) \in (x_1, x_2) \circ (y_1, y_2) = \{(a, b) \mid a \in x_1 \cdot \widehat{x_2}(y_1), b \in x_2 * y_2\},$$

which implies that $z_1 \in x_1 \cdot \widehat{x_2}(y_1)$ and $z_2 \in x_2 * y_2$. Since $z_1 \in x_1 \cdot \widehat{x_2}(y_1)$, we get $x_1 \in z_1 \cdot \widehat{x_2}(y_1)^{-1}$ or $x_1 \in z_1 \cdot \widehat{x_2}(y_1^{-1})$. Since $z_2 \in x_2 * y_2$, $x_2 \in z_2 * y_2^{-1}$. Therefore, $\gamma_B^*(x_2) = \gamma_B^*(z_2) \otimes \gamma_B^*(y_2^{-1})$ and so $\widehat{\gamma_B^*(x_2)} = \widehat{\gamma_B^*(z_2) \otimes \gamma_B^*(y_2^{-1})} = \widehat{\gamma_B^*(z_2)} \cdot \widehat{\gamma_B^*(y_2^{-1})}$ or $\widehat{x_2} = \widehat{z_2}\widehat{y_2^{-1}}$. Therefore, we get $x_1 \in z_1 \cdot \widehat{z_2}\widehat{y_2^{-1}}(y_1^{-1})$. Now, we have

$$(x_1, x_2) \in \{(a, b) \mid a \in z_1 \cdot \widehat{z_2}\widehat{y_2^{-1}}(y_1^{-1}), b \in z_2 * y_2^{-1}\}$$

or $(x_1, x_2) \in (z_1, z_2) \circ (y_1, y_2)^{-I}$. On the other hand, we have

$$(x_1, x_2)^{-I} \circ (z_1, z_2) = (\widehat{x_2^{-1}}(x_1^{-1}), x_2^{-1}) \circ (z_1, z_2))$$
$$= \{(a,, b) \mid a \in \widehat{x_2^{-1}}(x_1^{-1}) \cdot \widehat{x_2^{-1}}(z_1), b \in x_2^{-1} * z_2)\}$$
$$= \{(a, b) \mid a \in \widehat{x_2^{-1}}(x_1^{-1} \cdot z_1), b \in x_2^{-1} * z_2\}.$$

Since $z_1 \in x_1 \cdot \widehat{x_2}(y_1)$, $\widehat{x_2}(y_1) \in x_1^{-1} \cdot z_1$. Hence, $y_1 \in \widehat{x_2^{-1}}(x_1^{-1} \cdot z_1)$. Therefore, $(y_1, y_2) \in (x_1, x_2)^{-I} \circ (z_1, z_2)$. $\blacksquare$

Therefore, $A \times B$ equipped with the semidirect hyperproduct becomes a polygroup which we denote $A \widehat{\times} B$.

Theorem 3.5.16. *Let A and B be two polygroups, γ_A^* and γ_B^* be γ^*-relations on A and B respectively. If we consider $\widehat{} : B/\gamma_B^* \longrightarrow AutA$. Then,*

$$\widehat{b}(\gamma_A^*(a)) = \gamma_A^*(\widehat{b}(a)) \text{ for all } a \in A, b \in B.$$

Proof. Suppose that $x \in \widehat{b}(\gamma_A^*(a))$. Then, there exists $y \in \gamma_A^*(a)$ such that $x = \widehat{b}(y)$. So, there exist $(z_1, z_2, \ldots, z_n) \in A^n$ and $\sigma \in \mathbb{S}_n$ such that $a \in \prod_{i=1}^{n} z_i$ and $y \in \prod_{i=1}^{n} z_{\sigma(i)}$, which implies that

$$\widehat{b}(a) \in \prod_{i=1}^{n} \widehat{b}(z_i) \text{ and } \widehat{b}(y) \in \prod_{i=1}^{n} \widehat{b}(z_{\sigma(i)}),$$

that is, $\widehat{b}(a) \ \gamma_A^* \ \widehat{b}(y)$ or $\widehat{b}(y) = x \in \gamma_A^*(\widehat{b}(a))$.

Conversely, suppose that $x \in \gamma_A^*(\widehat{b}(a))$. Then, $x\gamma_A^*\widehat{b}(a)$ and so there exist $(z_1, z_2, \ldots, z_n) \in A^n$ and $\sigma \in \mathbb{S}_n$ such that $x \in \prod_{i=1}^{n} z_i$ and $\widehat{b}(a) \in \prod_{i=1}^{n} z_{\sigma(i)}$, which yields

$$\widehat{b}^{-1}(x) \in \prod_{i=1}^{n} \widehat{b}^{-1}(z_i) \text{ and } a \in \prod_{i=1}^{n} \widehat{b}^{-1}(z_{\sigma(i)}).$$

Hence, $\widehat{b}^{-1}(x)\gamma^*a$ or $\widehat{b}^{-1}(x) \in \gamma_A^*(a)$. Therefore, $x \in \widehat{b}(\gamma_A^*(a))$ and this completes the proof. $\blacksquare$

3.6 Generalized permutations

According to [69] a generalized permutation is defined as follows (see also [147; 148]).

Definition 3.6.1. Let Ω be a non-empty set. A map $f : \Omega \to \mathcal{P}^*(\Omega)$ is called a *generalized permutation* on Ω if

$$\bigcup_{\omega \in \Omega} f(\omega) = f(\Omega) = \Omega,$$

where $\mathcal{P}^*(\Omega)$ is the set of all non-empty subsets of Ω. We write

$$f = \begin{pmatrix} x \\ f(x) \end{pmatrix},$$

for the generalized permutation f. Denote M_Ω the set of all the generalized permutations on Ω.

Proposition 3.6.2. *Let Θ be a one to one function from a set Ω_1 onto a set Ω_2. For a generalized permutation f on Ω_1, we define a function $\Theta(f)$ on Ω_2 by the formula*

$$\Theta(f)(y) = \Theta(f(\Theta^{-1}(y))), \text{ for all } y \in \Omega_2.$$

Then, $\Theta(f)$ is a generalized permutation on Ω_2.

Proof. For every $y \in \Omega_2$, we have $\Theta^{-1}(y) \in \Omega_1$ which implies that $f(\Theta^{-1}(y)) \subseteq \Omega_1$, and so $\Theta(f(\Theta^{-1}(y))) \subseteq \Omega_2$. Furthermore,

$$\begin{aligned}
\bigcup_{y \in \Omega_2} \Theta(f)(y) &= \bigcup_{y \in \Omega_2} \Theta\Big(f(\Theta^{-1}(y))\Big) \\
&= \Theta\left(\bigcup_{y \in \Omega_2} f(\Theta^{-1}(y))\right) \\
&= \Theta(\Omega_1) = \Omega_2. \blacksquare
\end{aligned}$$

Definition 3.6.3. Let $f_1, f_2 \in M_\Omega$. We say that f_1 is a *subpermutation* of f_2 , or f_2 contains f_1, and write $f_1 \subseteq f_2$, if $f_1(x) \subseteq f_2(x)$ for every x in Ω. The mapping g for which $g(x) = \Omega$ for all $x \in \Omega$, is called the *universal generalized permutation* and contains all the elements of M_Ω.

Every map $f : \Omega \longrightarrow \mathcal{P}^*(\Omega)$ which contains a generalized permutation is itself a generalized permutation.

The map $i : \Omega \longrightarrow \mathcal{P}^*(\Omega)$ where $i(x) = \{x\} =: x$ for all $x \in \Omega$, is a generalized permutation.

We can define operation $\circ$, the usual composition on M_Ω, i.e., if $f, g \in M_\Omega$, then

$$f \circ g(x) = \bigcup_{y \in g(x)} f(y), \text{ for all } x \in \Omega.$$

Now, we define a hyperoperation $*$ on M_Ω as follows:

Definition 3.6.4. Let $\begin{pmatrix} x \\ f(x) \end{pmatrix}$ and $\begin{pmatrix} x \\ g(x) \end{pmatrix}$ be two elements of M_Ω. We define $* : M_\Omega \times M_\Omega \to \mathcal{P}^*(M_\Omega)$ by setting

$$\begin{pmatrix} x \\ g(x) \end{pmatrix} * \begin{pmatrix} x \\ f(x) \end{pmatrix} = \left\{ \begin{pmatrix} x \\ h(x) \end{pmatrix} \mid h \subseteq f \circ g, \bigcup_{x \in \Omega} h(x) = \Omega \right\}.$$

The generalized permutation $\begin{pmatrix} x \\ i(x) \end{pmatrix}$ serves as a scalar element, since

$$\left(\begin{matrix} x \\ f(x) \end{matrix}\right) \in \left(\begin{matrix} x \\ f(x) \end{matrix}\right) * \left(\begin{matrix} x \\ i(x) \end{matrix}\right) = \left(\begin{matrix} x \\ i(x) \end{matrix}\right) * \left(\begin{matrix} x \\ f(x) \end{matrix}\right).$$

For $f \in M_\Omega$ the inverse of f which is denoted by $\overline{f}$, is the generalized permutation defined as follows: $\overline{f}(y) = \{x \in \Omega \mid y \in f(x)\}$. It is clear that

$$\left(\begin{matrix} x \\ i(x) \end{matrix}\right) \in \left(\begin{matrix} x \\ f(x) \end{matrix}\right) * \left(\begin{matrix} x \\ \overline{f}(x) \end{matrix}\right) \cap \left(\begin{matrix} x \\ \overline{f}(x) \end{matrix}\right) * \left(\begin{matrix} x \\ f(x) \end{matrix}\right).$$

Proposition 3.6.5. *The hyperoperation $*$ is associative.*

Proof. Since $\circ$ is associative, it follows that $*$ is associative. ∎

Definition 3.6.6. $< P, \cdot, e, ^{-1} >$ is called a *poly-monoid* if the following conditions hold:

(1) $(x \cdot y) \cdot z = x \cdot (y \cdot z)$, for all $x, y, z \in P$,
(2) $x \in e \cdot x = x \cdot e$, for all $x \in P$,
(3) $e \in x \cdot x^{-1} \cap x^{-1} \cdot x$, for all $x \in P$.

Corollary 3.6.7. $< M_\Omega, *, i, ^- >$ *is a poly-monoid.*

Definition 3.6.8. Let $\mathcal{M} = < P, \cdot, e, ^{-1} >$ be a polygroup and M_Ω be the set of all generalized permutations on the non-empty set Ω. A mapping $\psi : P \to M_\Omega$ with the properties $\psi(x \cdot y) = \psi(x) * \psi(y)$ and $\psi(x^{-1}) = \overline{\psi}(x)$, for all $x, y \in P$, is called a *representation* of P by generalized permutations.

3.7 Permutation polygroups

Permutation polygroups are studied by Davvaz [49]. In this section, by using the concept of generalized permutation, we define permutation polygroups and some concepts related to it. In particular, we introduce a generalization of Cayley's theorem.

Definition 3.7.1 Let $\mathcal{M} = < P, \cdot, e, ^{-1} >$ be a polygroup and Ω be a non-empty set. A map $f : \Omega \times P \to \mathcal{P}^*(\Omega)$ is called an *action* of P on Ω if the following axioms hold:

(1) $f(\omega, e) = \{\omega\} = \omega$, for all $\omega \in \Omega$,
(2) $f(f(\omega, g), h) = \bigcup_{\alpha \in g.h} f(\omega, \alpha)$, for all $g, h \in P$ and $\omega \in \Omega$,
(3) $\bigcup_{\omega \in \Omega} f(\omega, g) = \Omega$, for all $g \in P$,
(4) $\forall g \in P, \alpha \in f(\beta, g) \Rightarrow \beta \in f(\alpha, g^{-1})$.

From the second condition, we obtain

$$\bigcup_{\omega_0 \in f(\omega, g)} f(\omega_0, h) = \bigcup_{\alpha \in g \cdot h} f(\omega, \alpha).$$

For $\omega \in \Omega$, we write $\omega^g := f(\omega, g)$. Then, we have

(1) $\omega^e = \omega$,

(2) $(\omega^g)^h = \omega^{g \cdot h}$, where $A^g = \bigcup_{\alpha \in A} \alpha^g$ and $\omega^B = \bigcup_{g \in B} \omega^g$, for all $A \subseteq \Omega$, $B \subseteq P$,

(3) $\bigcup_{\omega \in \Omega} \omega^g = \Omega$,

(4) $\forall g \in P, \ \alpha \in \beta^g \Rightarrow \beta \in \alpha^{g^{-1}}$.

In this case, we say that P is a *permutation polygroup* on a set Ω and it is said that P *acts on* Ω.

It is easy to see that if P is a permutation polygroup on two sets Ω_1 and Ω_2, then P is a permutation polygroup on the set $\Omega_1 \times \Omega_2$ with the action defined by $(\omega_1, \omega_2)^g = \{(a, b) \mid a \in \omega_1^g, b \in \omega_2^g\}$, for all $(\omega_1, \omega_2) \in \Omega_1 \times \Omega_2$ and $g \in P$.

The polygroup P acts on itself as a permutation polygroup if we define $x^g = x \cdot g$ or $x^g = g^{-1} \cdot x$, for all $x, g \in P$.

Proposition 3.7.2. *Let N be a normal subpolygroup of a polygroup P. Let Ω denote the set of all right cosets Nx, where $x \in P$, and we define*

$$(Nx)^g = \{Nz \mid z \in Nxg\}, \text{ for all } g \in P.$$

Then, P is a permutation polygroup on Ω.

Proof. It is easy to see that $(Nx)^e = Nx$. Now, if $g, h \in P$, then

$$((Nx)^g)^h = (\{Nz \mid z \in Nxg\})^h$$

$$= \bigcup_{z \in Nxg} \{Nt \mid t \in Nzh\}$$

$$= \{Nt \mid t \in Nxgh\}$$

$$= \bigcup_{\alpha \in g \cdot h} \{Nt \mid t \in Nx\alpha\}$$

$$= \bigcup_{\alpha \in g \cdot h} (Nx)^\alpha$$

$$= (Nx)^{g \cdot h}.$$

Therefore, the second condition of Definition 3.7.1 is satisfied. Now, we prove that $\bigcup_{Nx \in \Omega} (Nx)^g = \Omega$. Suppose that $Ny \in \Omega$, where $y \in P$. Since P is a hypergroup, there exists $a \in P$ such that $y \in ag$ which implies that $y \in Nag$, and so $Ny \in (Na)^g$. Therefore, $Ny \in \bigcup_{Nx \in \Omega} (Nx)^g$. Now, we show that

$$Nx \in (Ny)^g \Rightarrow Ny \in (Nx)^{g^{-1}}.$$

In order to prove this, we observe that since $Nx \in \{Nz \mid z \in Nyg\}$, there exists $z_0 \in Nyg$ such that $Nx = Nz_0$. From $z_0 \in Nyg$, we obtain $g \in y^{-1}Nz_0$. Hence, $y^{-1} \in gz_0^{-1}N$ and so $y \in Nz_0g^{-1}$. Therefore, $y \in Nxg^{-1}$ which implies that $Ny \in (Nx)^{g^{-1}}$. ∎

Definition 3.7.3. Let P be a polygroup and P acts on Ω. The *kernel* of action is defined as follows:

$$H = \{g \in P \mid \omega^g = \{\omega\}, \text{ for all } \omega \in \Omega\}.$$

Theorem 3.7.4. (Generalization of Cayley's Theorem). *Let P be a polygroup acting on a non-empty finite set Ω such that the action is faithful. Then, there is a subset of M_Ω which is a polygroup under the induced action of P and is isomorph to P.*

Proof. We define the subset S_Ω of M_Ω as follows:

$$S_\Omega = \left\{ \begin{pmatrix} \alpha_1 \; \alpha_2 \; \cdots \; \alpha_{|\Omega|} \\ \alpha_1^g \; \alpha_2^g \; \cdots \; \alpha_{|\Omega|}^g \end{pmatrix} \mid g \in P \right\}.$$

The hyperoperation $\circ$ on S_Ω is defined as follows:

$$\begin{pmatrix} \alpha_1 \; \alpha_2 \; \cdots \; \alpha_{|\Omega|} \\ \alpha_1^g \; \alpha_2^g \; \cdots \; \alpha_{|\Omega|}^g \end{pmatrix} \circ \begin{pmatrix} \alpha_1 \; \alpha_2 \; \cdots \; \alpha_{|\Omega|} \\ \alpha_1^h \; \alpha_2^h \; \cdots \; \alpha_{|\Omega|}^h \end{pmatrix}$$

$$= \left\{ \begin{pmatrix} \alpha_1 \; \alpha_2 \; \cdots \; \alpha_{|\Omega|} \\ \alpha_1^f \; \alpha_2^f \; \cdots \; \alpha_{|\Omega|}^f \end{pmatrix} \mid f \in g \cdot h \right\}.$$

Then, $< S_\Omega, \circ, i, ^{-I} >$ is a polygroup, where

$$\begin{pmatrix} \alpha_1 \; \alpha_2 \; \cdots \; \alpha_{|\Omega|} \\ \alpha_1^g \; \alpha_2^g \; \cdots \; \alpha_{|\Omega|}^g \end{pmatrix}^{-I} = \begin{pmatrix} \alpha_1 & \alpha_2 & \cdots & \alpha_{|\Omega|} \\ \alpha_1^{g^{-1}} & \alpha_2^{g^{-1}} & \cdots & \alpha_{|\Omega|}^{g^{-1}} \end{pmatrix}.$$

Now, we define $\phi : P \longrightarrow S_\Omega$ by

$$\phi(g) = \begin{pmatrix} \alpha_1 \; \alpha_2 \; \cdots \; \alpha_{|\Omega|} \\ \alpha_1^g \; \alpha_2^g \; \cdots \; \alpha_{|\Omega|}^g \end{pmatrix}.$$

It is easy to see that ϕ is well defined, one to one and onto. Moreover, ϕ is a strong homomorphism because, for every $g, h \in P$, we have

$$\phi(g \cdot h) = \{\phi(f) \mid f \in g \cdot h\}$$

$$= \left\{ \begin{pmatrix} \alpha_1 & \alpha_2 & \cdots & \alpha_{|\Omega|} \\ \alpha_1^f & \alpha_2^f & \cdots & \alpha_{|\Omega|}^f \end{pmatrix} \mid f \in g \cdot h \right\}$$

$$= \phi(g) \circ \phi(h).$$

Therefore, $P \cong S_\Omega$. ∎

Note that the above theorem is true when Ω is infinite.

Definition 3.7.5. Let P be a polygroup acting on non-empty sets Ω_1 and Ω_2. A map $\Theta : \Omega_1 \to \Omega_2$ is called a *P-map* if $\Theta(x^g) = \Theta(x)^g$, for all $g \in P$ and $x \in \Omega_1$.

If Θ is also a one to one correspondence, then Θ is called a *P-isomorphism* and Ω_1, Ω_2 are called *isomorphic*.

Proposition 3.7.6. *If P is a polygroup and λ, ρ are the left and the right regular representations of P, i.e., $\lambda_g : x \longrightarrow g^{-1} \cdot x$, $\rho_g : x \longrightarrow x \cdot g$, then (P, λ) and (P, ρ) are isomorphic.*

Proof. We define $\Theta : P \longrightarrow P$ by $\Theta(x) = x^{-1}$. Then,

$$\Theta(\rho_g x) = \Theta(x \cdot g) = \{\Theta(y) \mid y \in x \cdot g\} = \{y^{-1} \mid y \in x \cdot g\}$$
$$= (x \cdot g)^{-1} = g^{-1} \cdot x^{-1} = g^{-1} \cdot \Theta(x) = \lambda_g \Theta(x).$$

Since Θ is one to one and onto, so it is a *P*-isomorphism. ∎

Definition 3.7.7. Let P be a permutation polygroup on a non-empty set Ω. For $\alpha, \beta \in \Omega$ we define $\sim$ by

$$\alpha \sim \beta \text{ if and only if } \alpha \in \beta^g, \text{ for some } g \in P.$$

Lemma 3.7.8. *The relation defined above is an equivalence relation on Ω.*

Now, let

$$\Omega = \bigcup_{\alpha \in I} \triangle(\alpha)$$

be the partition of Ω with respect to this relation. Then, the sets $\triangle(\alpha), \alpha \in I$, are called *orbits* of P on Ω.

Definition 3.7.9. Let P be a permutation polygroup on a non-empty set Ω. If P has only one orbit, i.e., if $\alpha \sim \beta$ for every $\alpha, \beta \in \Omega$, we say that the polygroup is *transitive* on Ω. If $|I| > 1$, we say that P is *intransitive*.

Theorem 3.7.10. *Let P be a permutation polygroup on a non-empty set Ω. Then, the polygroup P is transitive on every orbit.*

Proof. Let $\triangle(\alpha)$ be the orbit containing $\alpha \in \Omega$. Clearly, for the set $\triangle(\alpha)$, conditions (1), (2) and (4) of Definition 3.7.1 hold. Therefore, we prove the third condition of Definition 3.7.1, i.e.,

$$\bigcup_{\omega \in \triangle(\alpha)} \omega^g = \triangle(\alpha), \text{ for all } g \in P.$$

In order to prove this, suppose that $\beta \in \displaystyle\bigcup_{\omega \in \triangle(\alpha)} \omega^g$. Then, $\beta \in \omega_0^g$ for some $\omega_0 \in \triangle(\alpha)$. So, $\beta \sim \omega_0$ and $\omega_0 \sim \alpha$ which imply that $\beta \sim \alpha$ or $\beta \in \triangle(\alpha)$. Therefore, $\displaystyle\bigcup_{\omega \in \triangle(\alpha)} \omega^g \subseteq \triangle(\alpha)$.

If $\triangle(\alpha_i), i \in I$ are all the disjoint orbits, then

$$\bigcup_{\omega \in \triangle(\alpha_i)} \omega^g \subseteq \triangle(\alpha_i), \text{ for all } i \in I.$$

But, $\displaystyle\bigcup_{\omega \in \Omega} \triangle(\alpha_i) = \Omega$ and $\triangle(\alpha_i) \cap \triangle(\alpha_j) = \emptyset$, for all $i \neq j$. So, $\displaystyle\bigcup_{\omega \in \triangle(\alpha_i)} \omega^g = \triangle(\alpha_i)$, for all $i \in I$. Therefore, P acts on $\triangle(\alpha)$.

Now, suppose that $\beta_1, \beta_2 \in \triangle(\alpha)$. Then, $\beta_1 \in \alpha^g, \beta_2 \in \alpha^h$ for some $g, h \in P$. By the forth condition of Definition 3.7.1, we have $\alpha \in \beta_2^{h^{-1}}$ and so

$$\beta_1 \in (\beta_2^{h^{-1}})^g = \beta_2^{h^{-1}g}.$$

Therefore, there exists $x \in h^{-1}g$ such that $\beta_1 \in \beta_2^x$. ∎

Corollary 3.7.11. *Let P be a permutation polygroup on a finite set Ω. If $\triangle$ is an orbit of P such that $|\Omega| = |\triangle|$, then P is transitive on Ω.*

Definition 3.7.12. Let P be a permutation polygroup on a non-empty set Ω and let $\omega \in \Omega$. The set

$$P_\omega = \left\{ g \in P \mid \omega^g = \omega^{g^{-1}} = \{\omega\} \right\} \subseteq P$$

is called the *stabilizer* of ω.

Corollary 3.7.13. *The stabilizer P_ω is a subpolgroup of P for each $\omega \in \Omega$.*

Definition 3.7.14. Let P be a permutation polygroup on a non-empty set Ω and $\omega_1, \cdots, \omega_k \in \Omega$. Then, the stabilizer $P_{\omega_1, \cdots, \omega_k}$ is the subpolygroup

$$P_{\omega_1,\cdots,\omega_k} = \left\{ g \in P \mid \omega_i^g = \omega_i^{g^{-1}} = \{\omega_i\}, \text{ for all } i = 1,\ldots,k \right\}.$$

This may be phrased also by

$$P_{\omega_1,\ldots,\omega_k} = \bigcap_{i=1}^{k} P_{\omega_i}.$$

It follows that if P acts on Ω and $\omega_1, \omega_2 \in \Omega$, then P_{ω_1} and P_{ω_2} act on Ω and

$$(P_{\omega_1})_{\omega_2} = P_{\omega_1,\omega_2} = (P_{\omega_2})_{\omega_1}.$$

Proposition 3.7.15. *Let N be a normal subpolygroup of a polygroup P. If Ω is the set of all the right cosets of N in P, then P acts on Ω and the kernel of this action is*

$$H = \bigcap_{x \in P} x^{-1} N x.$$

Proof. Suppose that $g \in H$. Then,

$$(Nx)^g = \{Nz \mid z \in Nxg\} = \{Nx\}.$$

From $z \in Nxg$, we obtain $g \in x^{-1}Nz = x^{-1}Nx$. Thus, $g \in \bigcap_{x \in P} x^{-1}Nx$.

Now, let $g \in \bigcap_{x \in P} x^{-1}Nx$. Then, $g \in x^{-1}Nx$, for every $x \in P$. We have

$$\begin{aligned}
(Nx)^g &= \{Nz \mid z \in Nxg\} \\
&\subseteq \{Nz \mid z \in Nxx^{-1}Nx\} \\
&\subseteq \{Nz \mid z \in NxN\} \\
&= \{Nz \mid z \in NNx\} \\
&= \{Nz \mid z \in Nx\} \\
&= \{Nx\}.
\end{aligned}$$

Therefore, $g \in H$. ∎

3.8   Representation of polygroups

Davvaz and Poursalavati studied the concept of representation of polygroups [59]. They introduced matrix representations of polygroups over hyperrings and shown such representations induce representations of the fundamental group over the corresponding fundamental ring.

A *hyperring* in the general sense is the largest class of multivalued systems that satisfies the ring-like axioms: $(R, +, \cdot)$ is a hyperring if $(R, +)$ is

a hypergroup, " $\cdot$ " is associative hyperoperation and the distributive laws
$$x \cdot (y + z) = x \cdot y + x \cdot z \text{ and } (x + y) \cdot z = x \cdot z + y \cdot z$$
are satisfied for every $x, y, z \in R$.

$(R, +, \cdot)$ is called a *semihyperring* if " $+$ ", " $\cdot$ " are associative hyperoperations, where " $\cdot$ " is distributive with respect to " $+$ ".

There are different notions of hyperrings. If only the addition " $+$ " is a hyperoperatiopn and the multiplication " $\cdot$ " is a usual operation, then R is called an *additive hyperring*. A special case of this type is the hyperring introduced by Krasner [95]. The second type of a hyperring was introduced by Rota [126]. The multiplication is a hyperoperation, while the addition is an operation, that is called *multiplication hyperring*. The monograph [56] is devoted especially to the study of hyperring theory.

In this section, we introduced the representation of polygroups by hypermatrices, and we introduced a polystructure on matrices. Also, we obtain some results in this connection.

A *hypermatrix* is a matrix with entries from a semihyperring. The hyperproduct of two hypermatrices $(a_{ij}), (b_{ij})$ which are of type $m \times n$ and $n \times r$ respectively, is defined in the usual manner
$$(a_{ij})(b_{ij}) = \left\{ (c_{ij}) \mid c_{ij} \in \sum_{k=1}^{n} a_{ik}b_{kj} \right\}.$$

One of the important problems concerning representation of polygroups is as follows:

For a given polygroup $\mathcal{P} = < P, \cdot, e, ^{-1} >$, find a semihyperring R with identity such that one gets a representation of $\mathcal{P}$ by hypermatrices with entries from R. Recall that if $M_R = \{(a_{ij}) | a_{ij} \in R\}$, then a map $T : P \longrightarrow M_R$ is called a *representation* if

(1) $T(x_1 \cdot x_2) = \{T(x) \mid x \in x_1 \cdot x_2\} = T(x_1)T(x_2)$, for all $x_1, x_2 \in P$,
(2) $T(e) = I$, where I is the identity matrix.

If instead of the first condition, we have the condition $T(x_1 \cdot x_2) \subseteq T(x_1)T(x_2)$, for all $x_1, x_2 \in P$, then T is called an *inclusion representation*.

Example 3.8.1. Suppose that the multiplication table for the polygroup $\mathcal{P} = (P, \cdot, e, ^{-1})$, where $P = \{e, a, b\}$ is

$\cdot$	e	a	b
e	e	a	b
a	a	$\{e, b\}$	$\{a, b\}$
b	b	$\{a, b\}$	$\{e, a\}$

In $\mathbb{Z}_3$, we define a hyperoperation $\oplus$ as follows: $1 \oplus 1 = \{0, 2\}$, $2 \oplus 2 = \{0, 1\}$, $1 \oplus 2 = 2 \oplus 1 = \{1, 2\}$ and $\oplus = +$ be the usual sum for the other cases, and let $\odot$ be the usual product in $\mathbb{Z}_3$. One can see that $(\mathbb{Z}_3, \oplus, \odot)$ is a semihyperring. Then, the map $T : P \longrightarrow M_R$ with

$$T(e) = \begin{pmatrix} 1 & 0 & 0 \\ 0 & 1 & 0 \\ 0 & 0 & 1 \end{pmatrix}, \quad T(a) = \begin{pmatrix} 1 & 0 & 1 \\ 0 & 1 & 0 \\ 0 & 0 & 1 \end{pmatrix}, \quad T(b) = \begin{pmatrix} 1 & 0 & 2 \\ 0 & 1 & 0 \\ 0 & 0 & 1 \end{pmatrix}$$

is a representation of the polygroup P.

Generally, if we choose i_0, j_0, $i_0 \neq j_0$, $0 \leq i_0$, $j_0 \leq n$ and then put $T(e) = I_n$, $T(a) = A_n$ and $T(b) = B_n$ where

$$A_n = (a_{ij}) \text{ with } \begin{cases} a_{ii} = 1 & \text{for } i = 1, \cdots, n \\ a_{i_0 j_0} = 1 \\ a_{ij} = 0 & \text{otherwise.} \end{cases}$$

$$B_n = (b_{ij}) \text{ with } \begin{cases} b_{ij} = a_{ij} & \text{if } i \neq i_0, \ j \neq j_0 \\ b_{i_0 j_0} = 2. \end{cases}$$

Then, T is a representation of P.

Example 3.8.2. Suppose that $< P_1, \cdot, e_1, ^{-1} >$ and $< P_2, *, e_2, ^{-1} >$ are two polygroups. We know $< P_1 \times P_2, \circ, E, ^{-I} >$ is a polygroup, where $(x_1, y_1) \circ (x_2, y_2) = \{(x, y) \mid x \in x_1 \cdot x_2, y \in y_1 * y_2\}$, $E = (e_1, e_2)$, $(x, y)^{-I} = (x^{-1}, y^{-1})$. Now, if $T_1 : P_1 \longrightarrow M_R$ and $T_2 : P_2 \longrightarrow M_R$ are two representations of P_1 and P_2 respectively, then we have the following representation for $P_1 \times P_2$:

$$T_1 \times T_2 : P_1 \times P_2 \longrightarrow M_R,$$
$$T_1 \times T_2(x, y) = \begin{bmatrix} T_1(x) & 0 \\ 0 & T_2(y) \end{bmatrix}.$$

Proposition 3.8.3. *Let $\mathcal{A} =< A, \cdot, e, ^{-1} >$ and $\mathcal{B} =< B, \cdot, e, ^{-1} >$ be two polygroups. Let T be a representation of $\mathcal{B}$. Then, $\varphi : \mathcal{A}[\mathcal{B}] \longrightarrow M_R$ where*

$$\varphi(x) = \begin{cases} T(x) & \text{if } x \in B \\ I_n & \text{if } x \in A \end{cases}$$

is a representation of $\mathcal{A}[\mathcal{B}]$.

Proof. If $x, y \in B$ and $y = x^{-1}$, then $\varphi(x)\varphi(y) = T(x)T(y)$ and $\varphi(x * y) = \varphi(x \cdot y \cup A) = \varphi(x \cdot y) \cup \varphi(A) = \varphi(x \cdot y) \cup \{I_n\}$. Since $e \in x \cdot y$, $\varphi(e) \in \varphi(x \cdot y)$ and so $I_n \in \varphi(x \cdot y)$. Therefore, $\varphi(x * y) = \varphi(x)\varphi(y)$. Other cases are obvious. ■

We recall the following definition from [56; 147].

Definition 3.8.4. Let $(R, +, \cdot)$ be a hyperring. We define the relation Γ as follows:

$a\Gamma b$ if and only if $\{a, b\} \subseteq u$, where u is a finite sum of finite products of elements of R.

We denote the transitive closure of Γ by Γ^*. The equivalence relation Γ^* is called the *fundamental equivalence relation* in R. We denote the equivalence class of the element a (also called the *fundamental class of a*) by $\Gamma^*(a)$.

According to the distributive law, every set which is the value of a polynomial in elements of R is a subset of a sum of products in R.

Let $\mathcal{U}$ be the set of all finite sums of products of elements of R. We can rewrite the definition of Γ^* on R as follows:

$a \ \Gamma^* \ b$ if and only if there exist $z_1, \ldots, z_{n+1} \in R$ with $z_1 = a$, $z_{n+1} = b$ and $u_1, \ldots, u_n \in \mathcal{U}$ such that $\{z_i, z_{i+1}\} \subseteq u_i$ for $i \in \{1, \ldots, n\}$.

Theorem 3.8.5. *Let $(R, +, \cdot)$ be a hyperring. Then, the relation Γ^* is the smallest equivalence relation on R such that the quotient R/Γ^* is a ring.*

R/Γ^* is called the *fundamental ring*.

Proof. First, we prove that R/Γ^* is a ring. The product $\odot$ and the sum $\oplus$ in R/Γ^* are defined as follows:

$$\Gamma^*(a) \oplus \Gamma^*(b) = \{\Gamma^*(c) \mid c \in \Gamma^*(a) + \Gamma^*(b)\},$$
$$\Gamma^*(a) \odot \Gamma^*(b) = \{\Gamma^*(d) \mid d \in \Gamma^*(a) \cdot \Gamma^*(b)\}.$$

Let $a' \in \Gamma^*(a), b' \in \Gamma^*(b)$. Hence,

$a' \ \Gamma^* \ a$ implies that there exist $x_1, \ldots, x_{m+1}$ with $x_1 = a'$, $x_{m+1} = a$ and $u_1, \ldots, u_m \in \mathcal{U}$ such that $\{x_i, x_{i+1}\} \subseteq u_i$ for $i \in \{1, \ldots, m\}$;

$b' \ \Gamma^* \ b$ implies that there exist $y_1, \ldots, y_{n+1}$ with $y_1 = b'$, $y_{n+1} = b$ and $v_1, \ldots, v_n \in \mathcal{U}$ such that $\{y_j, y_{j+1}\} \subseteq v_j$ for $j \in \{1, \ldots, n\}$.

We obtain

$$\{x_i, x_{i+1}\} + y_1 \subseteq u_i + v_1, \ i \in \{1, \ldots, m-1\}$$
$$x_{m+1} + \{y_j, y_{j+1}\} \subseteq u_m + v_j, \ j \in \{1, \ldots, n\}.$$

The sets $u_i + v_1 = t_i$, $i \in \{1, \ldots, m-1\}$ and $u_m + v_j = t_{m+j-1}$, $j \in \{1, \ldots, n\}$ are elements of $\mathcal{U}$. We choose the elements $z_1, \ldots, z_{m+n}$ such that

$$z_i \in x_i + y_1, i \in \{1, \ldots, m\}$$

and

$$z_{m+j} \in x_{m+1} + y_{j+1}, j \in \{1, \ldots, n\}.$$

We obtain $\{z_k, z_{k+1}\} \subseteq t_k$, $k \in \{1, \ldots, m+n-1\}$. Hence, any element $z_1 \in x_1 + y_1 = a' + b'$ is Γ^* equivalent to any element $z_{m+n} \in x_{m+1} + y_{n+1} = a + b$. Thus, $\Gamma^*(a) \oplus \Gamma^*(b)$ is singleton and we have

$$\Gamma^*(a) \oplus \Gamma^*(b) = \Gamma^*(c), \text{ for all } c \in \Gamma^*(a) + \Gamma^*(b).$$

According to the distributive law, we have $u \cdot v \in \mathcal{U}$ for all $u, v \in \mathcal{U}$. Similarly, we obtain $\Gamma^*(a) \odot \Gamma^*(b) = \Gamma^*(d)$ for all $d \in \Gamma^*(a) \cdot \Gamma^*(b)$. Therefore, it is immediate that R/Γ^* is a ring.

Now, let ρ be an equivalence relation in R, such that R/ρ is a ring and let $\rho(a)$ be the equivalence class of the element a. Then, $\rho(a) \oplus \rho(b)$ and $\rho(a) \odot \rho(b)$ are singletons for all $a, b \in R$, which means that for all $a, b \in R$, for all $c \in \rho(a) + \rho(b)$, for all $d \in \rho(a) \cdot \rho(b)$ we have

$$\rho(a) \oplus \rho(b) = \rho(c), \quad \rho(a) \odot \rho(b) = \rho(d).$$

The above equalities are called the *fundamental properties* in $(R/\rho, \oplus, \odot)$. Hence, for all $a, b \in R$ and $A \subseteq \rho(a)$, $B \subseteq \rho(b)$ we have

$$\rho(a) \oplus \rho(b) = \rho(a + b) = \rho(A + B)$$

and

$$\rho(a) \odot \rho(b) = \rho(a \cdot b) = \rho(A \cdot B).$$

By induction, we extend the above equalities to finite sums and products. Now, set $u \in \mathcal{U}$, which means that there exist the finite sets of indices J and I_j and the elements $x_i \in R$ such that:

$$u = \sum_{j \in J} \left(\prod_{i \in I_j} x_i \right).$$

For all I_j, the set $\prod_{i \in I_j} x_i$ is a subset of one class, say $\rho(a_j)$. Thus, for all $a \in \sum_{j \in J} a_j$ we have

$$u \subseteq \sum_{j \in J} \rho(a_j) = \rho \left(\sum_{j \in J} a_j \right) = \rho(a).$$

Therefore, for all $x, y \in R$, $x \, \Gamma \, y$ implies $x \, \rho \, y$, whence $x \, \Gamma^* \, y$ implies that $x \, \rho \, y$. Hence, for all $a \in R$, $\Gamma^*(a) \subseteq \rho(a)$, which means that Γ^* as the smallest equivalence relation in R such that the quotient R/Γ^* is a ring. ∎

Remark 3.8.6. If $u = \sum_{j \in J} \left(\prod_{i \in I_j} x_i \right) \in \mathcal{U}$, then for all $z \in u$,

$$\Gamma^*(u) = \oplus \sum_{j \in J} \left(\odot \prod_{i \in I_j} \Gamma^*(x_i) \right) = \Gamma^*(z),$$

where $\oplus \sum$ and $\odot \prod$ denote the sum and the product of classes.

In order to speak about canonical maps, we need the following notion:

Definition 3.8.7. Let R_1 and R_2 be two hyperrings. The map $f : R_1 \to R_2$ is called an *inclusion homomorphism* if for all $x, y \in R$, the following conditions hold:

$$f(x + y) \subseteq f(x) + f(y) \text{ and } f(x \cdot y) \subseteq f(x) \cdot f(y).$$

f is called a *strong homomorphism* if for all $x, y \in R$, we have

$$f(x + y) = f(x) + f(y) \text{ and } f(x \cdot y) = f(x) \cdot f(y).$$

Let R be a hyperring. We denote by $\beta_{\cdot}$ and β_{+} the following binary relations:

$x \beta_{\cdot} y$ if and only if there exist $z_1, \ldots, z_n \in R$ such that $\{x, y\} \subseteq z_1 \cdot \ldots \cdot z_n$

and

$x \beta_{+} y$ if and only if there exist $z_1, \ldots, z_n \in R$ such that $\{x, y\} \subseteq z_1 + \ldots + z_n$.

We denote the transitive closures of the relations $\beta_{\cdot}$ and β_{+} by $\beta_{\cdot}^*$ and β_{+}^*, and we call $\beta_{\cdot}^*$ and β_{+}^* the *fundamental equivalence relations* with respect to multiplication and addition, respectively. For all $a \in R$ we denote the corresponding equivalence classes of a by $\beta_{\cdot}^*(a)$ and $\beta_{+}^*(a)$ and we have

$$\beta_{\cdot}^*(a) \subseteq \Gamma^*(a), \ \beta_{+}^*(a) \subseteq \Gamma^*(a).$$

Let us consider the following canonical maps

$$\varphi_{\cdot} : R \longrightarrow R/\beta_{\cdot}^*, \ \varphi_{\cdot}(x) = \beta_{\cdot}^*(x),$$
$$\varphi_{+} : R \longrightarrow R/\beta_{+}^*, \ \varphi_{+}(x) = \beta_{+}^*(x),$$
$$\varphi^* : R \longrightarrow R/\Gamma^*, \ \varphi^*(x) = \Gamma^*(x).$$

We notice that the maps $\varphi_{+} : (R, +) \longrightarrow (R/\beta_{+}^*, \oplus)$, $\varphi_{\cdot} : (R, \cdot) \longrightarrow (R/\beta_{\cdot}^*, \odot)$, $\varphi^* : (R, +, \cdot) \longrightarrow (R/\Gamma^*, \oplus, \odot)$ are strong homomorphisms.

We denote by ω_{+}, ω^* the kernels of φ_{+}, φ^*, respectively. If $\bar{0}$ is the zero element of R/β_{+}^* or R/Γ^*, then

$$\omega_{+} = ker\varphi_{+} = \{x \in R : \varphi_{+}(x) = \bar{0}\},$$
$$\omega^* = ker\varphi^* = \{x \in R : \varphi^*(x) = \bar{0}\}.$$

We have $\omega_{+} \subseteq \omega^*$.

Theorem 3.8.8. *A necessary condition in order to have an inclusion representation T of polygroup $< P, \cdot, e,^{-1} >$ by $n \times n$ hypermatrices over the hyperring $(R, +, \cdot)$ is the following:*

- *For every fundamental class $\beta^*(x)$, there must exist elements $a_{ij} \in R$ $(i, j \in \{1, \ldots, n\})$ such that*

$$T(\beta^*(x)) \subseteq \Big\{ (a'_{ij}) \mid a'_{ij} \in \Gamma^*(a'_{ij}), \; i, j \in \{1, \ldots, n\} \Big\}.$$

Proof. Let $x \; \beta^* \; y$, for $x, y \in P$ and $T(x) = (x_{ij})$, $T(y) = (y_{ij})$, where $x_{ij}, y_{ij} \in R$ for all $i, j \in \{1, \ldots, n\}$. Then, there exist elements $z_1, \ldots, z_{r-1}$, $h_{q_1}, \ldots, h_{q_r} \in P$ and $Q_1, \ldots, Q_r$ finite sets of indices, such that setting $z_0 = x$, $z_r = y$, then

$$\{z_{v-1}, z_v\} \subseteq \prod_{p_v, q_v \in Q_v} h_{q_v}, \quad v \in \{1, \ldots, r\}.$$

Therefore, for all $v \in \{1, \ldots, r\}$, we obtain

$$\{T(z_{v-1}), T(z_v)\} \subseteq T \left(\prod_{p_v, q_v \in Q_v} h_{q_v} \right)$$
$$\subseteq \prod_{p_v, q_v \in Q_v} T(h_{q_v}).$$

In the ij entry of the hypermatrix

$$\prod_{p_v, q_v \in Q_v} h_{q_v}$$

there are sets which are the union of sets U_μ formed by sums of $n^{|Q_v|-1}$ products of Q_v elements each. Consequently, each of these sets U_μ belong to one Γ^* fundamental class. Furthermore, for every U_μ, there exists another U_λ such that $U_\mu \cap U_\lambda \neq \emptyset$. Since Γ^* is transitive, there exists an element $a_{v,ij} \in R$ such that for all ij entries $z_{v-1,ij}$ and $z_{v,ij}$ of the hypermatrices $T(z_{v-1})$ and $T(z_v)$, respectively, we have

$$\{z_{v-1,ij}, z_{v,ij}\} \subseteq \Gamma^*(a_{v,ij}), \text{ for all } v \in \{1, \ldots, r\}.$$

Again, from the transitivity of Γ^*, we have

$$\Gamma^*(a_{1,ij}) = \ldots = \Gamma^*(a_{v,ij}) := \Gamma^*(a_{ij}).$$

In particular, we obtain

$$\{x_{ij}, y_{ij}\} \subseteq \Gamma^*(a_{ij}), \text{ for all } i, j \in \{1, \ldots, n\}. \quad \blacksquare$$

Theorem 3.8.9. *Every representation $T(a) = (a_{ij})$ of a polygroup $< P, \cdot, e, ^{-1} >$ by $n \times n$ hypermatrices over the hyperring $(R, +, \cdot)$ induces a $n \times n$ representation T^* of the fundamental group P/β^* over the fundamental ring R/Γ^* by setting*

$$T^*(\beta^*(a)) = (\Gamma^*(a_{ij})), \text{ for all } \beta^*(a) \in P/\beta^*.$$

Proof. By using Theorem 3.8.8, it is straightforward. ∎

Let $\mathcal{P} = < P, \cdot, e, ^{-1} >$ be a polygroup. Two elements $x, y \in P$ are said to be *conjugate* if there exists an elements $z \in P$ such that $y \in z^{-1}xz$.

Theorem 3.8.10. *Let T be a representation of P over R of degree n and let I_n be the unit matrix over R/Γ^*. Then,*

(1) $T^*(\beta^*(e)) = I_n$;
(2) $T^*(\beta^*(x^{-1})) = (T^*(\beta^*(x)))^{-1}$, for all $x \in P$;
(3) If $x, y \in P$ are conjugate, then $T^*(\beta^*(x))$, $T^*(\beta^*(y))$ are conjugate.

Proof. (1) For every $x \in P$, we have $\beta^*(x) = \beta^*(x) \odot \beta^*(e) = \beta^*(e) \odot \beta^*(x)$ and so we obtain

$$T^*(\beta^*(x)) = T^*(\beta^*(x))T^*(\beta^*(e)) = T^*(\beta^*(e))T^*(\beta^*(x)).$$

Therefore, $T^*(\beta^*(e)) = I_n$.

(2) We have $e \in x \cdot x^{-1}$ and so $\beta^*(e) = \beta^*\ (x \cdot x^{-1}) = \beta^*(x) \odot \beta^*(x^{-1})$ which implies that

$$T^*(\beta^*(e)) = T^*(\beta^*(x) \odot \beta^*(x^{-1})) = T^*(\beta^*(x))T^*(\beta^*(x^{-1})).$$

So, $I_n = T^*(\beta^*(x))T^*(\beta^*(x^{-1}))$. Therefore, $T^*(\beta^*(x^{-1})) = (T^*(\beta^*(x))^{-1}$.

(3) The proof is similar to the proofs of (1) and (2). ∎

3.9 Polygroup hyperrings

Davvaz and Poursalavati introduced the notion of polygroup hyperrings as a generalization of group rings [59]. They established homomorphisms among various polygroup hyperrings.

Let $< P, \cdot, e, ^{-1} >$ be a finite polygroup, and $< R, +, 0, - >$ be a commutative polygroup and $(R, +, *)$ be a hyperring with scalar unit and zero element. Suppose that $R[P]$ is the set of all the functions on P with values in R, i.e.,

$$R[P] = \{f \mid f : P \longrightarrow R \text{ is a function}\}.$$

On $R[P]$ we consider the hyperoperations defined as follow:

$$f \oplus g = \{h \mid h(x) \in f(x) + g(x)\},$$

$$f \odot g = \left\{ h \mid h(z) \in \sum_{z \in x \cdot y} f(x) * g(y) \right\}.$$

We define the mapping $^{-I} : R[P] \longrightarrow R[P]$, where $f^{-I} : P \longrightarrow R$ is defined by $f^{-I}(p) = -f(p)$ for every $p \in P$ and let f_0 be the zero map. Our aim in the following lemma and theorem is to show that $R[P]$ is a hyperring with hyperoperations $\oplus$ and $\odot$.

Lemma 3.9.1. $< R[P], \oplus, f_0, ^{-I} >$ *is a polygroup.*

Proof. For every $f, g, h \in R[P]$, obviously we have $(f \oplus g) \oplus h = f \oplus (g \oplus h)$ and $f_0 \oplus f = f \oplus f_0 = f$. Now, let $f \in g \oplus h$. Then, for every $x \in P$ we have $f(x) \in g(x) + h(x)$ and so $g(x) \in f(x) - h(x)$ and $h(x) \in -g(x) + f(x)$. Therefore, $g \in f \oplus (-h)$ and $h \in (-g) \oplus f$ ∎

Theorem 3.9.2. $(R[P], \oplus, \odot)$ *is a hyperring.*

Proof. By Lemma 3.9.1, $< R[P], \oplus, f_0, ^{-I} >$ is a polygroup. Let $f_1, f_2, f_3 \in R[P]$. Then,

$$f_1 \odot (f_2 \odot f_3)$$

$$= f_1 \odot \left\{ f \mid f(z) \in \sum_{z \in x \cdot y} f_2(x) * f_3(y) \right\}$$

$$= \bigcup_f f_1 \odot f \text{ where the union is over } f \text{ with}$$

$$f(z) \in \sum_{z \in x \cdot y} f_2(x) * f_3(y)$$

$$= \bigcup_f \left\{ g \mid g(a) \in \sum_{a \in b \cdot c} f_1(b) * f(c) \right\}$$

$$= \left\{ g \mid g(a) \in \sum_{a \in b \cdot c} f_1(b) * \left(\sum_{c \in x \cdot y} f_2(x) * f_3(y) \right) \right\}$$

$$= \left\{ g \mid g(a) \in \sum_{a \in b \cdot c} \sum_{c \in x \cdot y} f_1(b) * (f_2(x) * f_3(y)) \right\}$$

$$= \left\{ g \mid g(a) \in \sum_{a \in b \cdot c} \sum_{c \in x \cdot y} (f_1(b) * f_2(x)) * f_3(y) \right\}$$

$$= \left\{ g \mid g(a) \in \sum_{a \in b \cdot (x \cdot y)} (f_1(b) * f_2(x)) * f_3(y) \right\}$$

$$= \left\{ g \mid g(a) \in \sum_{a \in (b \cdot x) \cdot y} (f_1(b) * f_2(x)) * f_3(y) \right\}$$

$$= \left\{ g \mid g(a) \in \sum_{a \in d \cdot y} \sum_{d \in b \cdot x} (f_1(b) * f_2(x)) * f_3(y) \right\}$$

$$= \bigcup_f f \odot f_3 \text{ where the union is over } f \text{ with}$$

$$f(z) \in \sum_{z \in x \cdot y} f_1(x) * f_2(y)$$

$$= (f_1 \odot f_2) \odot f_3.$$

Similarly, we have

$$f \odot (f_1 \oplus f_2) = \bigcup_{g \in f_1 \oplus f_2} f \odot g = \bigcup_{g \in f_1 \oplus f_2} \left\{ h \mid h(z) \in \sum_{z \in x \cdot y} f(x) * g(y) \right\}$$

$$= \left\{ h \mid h(z) \in \sum_{z \in x \cdot y} f(x) * (f_1(y) + f_2(y)) \right\}$$

$$= \left\{ h \mid h(z) \in \sum_{z \in x \cdot y} (f(x) * f_1(y)) + (f(x) * f_2(y)) \right\}$$

$$= \left\{ h \mid h(z) \in \left(\sum_{z \in x \cdot y} f(x) * f_1(y) \right) + \left(\sum_{z \in x \cdot y} f(x) * f_2(y) \right) \right\}$$

$$= \bigcup_{h_1, h_2} \{ h \mid h(z) \in h_1(z) + h_2(z) \}$$

where the union is over $h_1 \in f \odot f_1$ and $h_2 \in f \odot f_2$

$$= \bigcup_{h_1, h_2} h_1 \oplus h_2 = \bigcup_{h_1} \bigcup_{h_2} (h_1 \oplus h_2) = \bigcup_{h_1} \left(\bigcup_{h_2} h_1 \oplus h_2 \right)$$

$$= \bigcup_{h_2} (f \odot f_1) \oplus h_2 = (f \odot f_1) \oplus (f \odot f_2).$$

Hence, $f \odot (f_1 \oplus f_2) = (f \odot f_1) \oplus (f \odot f_2)$. Similarly it can be proved that $(f_1 \oplus f_2) \odot f = (f_1 \odot f) \oplus (f_2 \odot f)$. Consequently, $(R[P], \oplus, \odot)$ is a hyperring. ∎

Now that we constructed the hyperring $R[P]$ from R and P we will study relation between the polygroup P and the hyperring $R[P]$.

We define $E : R \longrightarrow R[P]$ by $r \mapsto E_r$, where $E_r : P \longrightarrow R$ is defined by

$$E_r(g) = \begin{cases} r \text{ if } g = e \\ 0 \text{ if } g \neq e. \end{cases}$$

It is clear that E is a one to one function and we have

$$E(r_1 + r_2) = E(r_1) \oplus E(r_2),$$
$$E(r_1 * r_2) = E(r_1) \odot E(r_2),$$
$$E(0) = E_0 := \text{zero function}.$$

Therefore, R is imbedded in $R[P]$.

If H is a subpolygroup of P, then we write

$$R\langle H \rangle = \{ f \in R[P] \mid \{ x \mid f(x) \neq 0 \} \subseteq H \}.$$

Then, there is a one to one polygroup homomorphism from $R\langle H \rangle$ to $R[P]$.

Proposition 3.9.3. *Let P_1 and P_2 be two polygroups and $\psi : P_1 \longrightarrow P_2$ be a mapping. Then, there exists an inclusion homomorphism of polygroups $\varphi : R[P_1] \longrightarrow R[P_2]$.*

Proof. We define $\varphi(f) = f \circ \psi$. Obviously, φ is well defined. If $h \in f_1 \oplus f_2$, then for every $x \in P_1$, we have $\varphi(h)(x) = h(\psi(x)) \in f_1(\psi(x)) + f_2(\psi(x))$ or $\varphi(h)(x) \in \varphi(f_1)(x) + \varphi(f_2(x))$ which implies that $\varphi(h) \in \varphi(f_1) \oplus \varphi(f_2)$ and so $\varphi(f_1 \oplus f_2) \subseteq \varphi(f_1) \oplus \varphi(f_2)$. ∎

Proposition 3.9.4. *Let $\psi : R \longrightarrow S$ be a surjective inclusion homomorphism of hyperrings and $T = \ker\psi$. Then, the mapping $\overline{\psi} : R[P] \longrightarrow S[P]$ denoted by $\overline{\psi}(f) = \psi \circ f$ is a surjective inclusion homomorphism whose kernel is $T[P]$.*

Proof. We have

$$
\begin{aligned}
\overline{\psi}(f_1 \oplus f_2) &= \psi \circ (f_1 \oplus f_2) = \psi \circ \{ f \mid f \in f_1 \oplus f_2 \} \\
&= \{ \psi \circ f \mid f \in f_1 \oplus f_2 \} \\
&= \{ h \mid h(p) = \psi(f(p)),\ f(p) \in f_1(p) + f_2(p), \forall p \in P \} \\
&= \{ h \mid h(p) \in \psi(f_1(p) + f_2(p)), \forall p \in P \} \\
&\subseteq \{ h \mid h(p) \in \psi(f_1(p)) + \psi(f_2(p)) \} = \psi \circ f_1 \oplus \psi \circ f_2,
\end{aligned}
$$

and so

$$
\begin{aligned}
\overline{\psi}(f_1 \odot f_2) &= \left\{ \overline{\psi}(f) \;\middle|\; f(x) \in \sum_{x \in y \cdot z} f_1(y) * f_2(z) \right\} \\
&= \left\{ \psi \circ f \;\middle|\; f(x) \in \sum_{x \in y \cdot z} f_1(y) * f_2(z) \right\} \\
&\subseteq \left\{ \psi \circ f \;\middle|\; \psi \circ f(x) \in \psi\left(\sum_{x \in y \cdot z} f_1(y) * f_2(z) \right) \right\} \\
&\subseteq \left\{ \psi \circ f \;\middle|\; \psi \circ f(x) \in \sum_{x \in y \cdot z} \psi(f_1(y)) * \psi(f_2(z)) \right\} \\
&= \overline{\psi}(f_1) \odot \overline{\psi}(f_2).
\end{aligned}
$$

Therefore, $\overline{\psi}$ is an inclusion homomorphism. Obviously, $\overline{\psi}$ is onto, and

$$
\begin{aligned}
\ker\overline{\psi} &= \{ f \in R[P] \mid \psi \circ f = f_0 \},\ \text{where } f_0 \text{ is the zero function} \\
&= \{ f \in R[P] \mid \psi(f(x)) = 0, \forall x \in P \} \\
&= \{ f \in R[P] \mid f(x) \in \ker\psi, \forall x \in P \} \\
&= \{ f \in R[P] \mid f(x) \in T, \forall x \in P \} = T[P]. \blacksquare
\end{aligned}
$$

Let Γ_1^* and Γ_2^* be the fundamental relations on R and $R[P]$, respectively. Let $\mathcal{U}_1$ and $\mathcal{U}_2$ denote the sets of all finite polynomials of elements of R and $R[P]$ over natural numbers, respectively. In the following theorem we will construct a homomorphism between R/Γ_1^* and $R[P]/\Gamma_2^*$.

Theorem 3.9.5. *There is a homomorphism* $g : R/\Gamma_1^* \longrightarrow R[P]/\Gamma_2^*$.

Proof. We define $g(\Gamma_1^*(r)) = \Gamma_2^*(E_r)$. First, we prove that g is well defined. We know $a \; \Gamma_1^* \; b$ if and only if there exist $x_1, \ldots, x_{m+1} \in R$; $u_1, \ldots, u_m \in \mathcal{U}_1$ with $x_1 = a$, $x_{m+1} = b$ such that $\{x_i, x_{i+1}\} \subseteq u_i$, $i = 1, \ldots, m$. Then, this implies $E(\{x_i, x_{i+1}\}) \subseteq E(u_i)$ or $\{E(x_i), E(x_{i+1})\} \subseteq E(u_i) \in \mathcal{U}_2$ and so $E(x_i)\Gamma_2^* E(x_{i+1})$, $i = 1, \ldots, m$. Therefore, $E(a)\Gamma_2^* E(b)$ that is to say $\Gamma_2^*(E_a) = \Gamma_2^*(E_b)$.

Now, we will show that g is a homomorphism. This is because

$$g(\Gamma_1^*(a) \oplus \Gamma_1^*(b)) = g(\Gamma_1^*(a + b)) = \Gamma_2^*(E(a + b))$$
$$= \Gamma_2^*(E(a) \oplus E(b)) = \Gamma_2^*(E(a)) \oplus \Gamma_2^*(E(b))$$
$$= g(\Gamma_1^*(a)) \oplus g(\Gamma_1^*(b)).$$

Similarly, we obtain $g(\Gamma_1^*(a) \odot \Gamma_1^*(b)) = g(\Gamma_1^*(a)) \odot g(\Gamma_1^*(b))$. ∎

Corollary 3.9.6. *The following diagram is commutative, i.e.,* $\varphi_2 E = g\varphi_1$, *where* φ_1 *and* φ_2 *are canonical maps.*

$$
\begin{array}{ccc}
R & \xrightarrow{\;\;E\;\;} & R[P] \\
\varphi_1 \downarrow & & \downarrow \varphi_2 \\
R/\Gamma_1^* & \xrightarrow{\;\;g\;\;} & R[P]/\Gamma_2^*
\end{array}
$$

3.10 Solvable polygroups

Aghabozorgi, Davvaz and Jafarpour in [2] introduced the concept of solvable polygroups. The purpose of this section is to provide a detailed structure description of derived subpolygroups of polygroups. We investigate the concept of perfect and solvable polygroups and we give some results in this respect. Finally, we discuss on τ-multi-semi-direct hyperproduct of polygroups.

Definition 3.10.1. Let H be a hypergroup. We define

(1) $[x, y]_r = \{h \in H \mid x \cdot y \cap y \cdot x \cdot h \neq \emptyset\}$;

(2) $[x, y]_l = \{h \in H \mid x \cdot y \cap h \cdot y \cdot x \neq \emptyset\}$;

(3) $[x, y] = [x, y]_r \cup [x, y]_l$.

From now on we call $[x, y]_r$, $[x, y]_l$ and $[x, y]$ *right commutator x and y, left commutator x and y* and *commutator x and y*, respectively. Also, we will denote $[H, H]_r$, $[H, H]_l$ and $[H, H]$ the set of all right commutators, left commutators and commutators, respectively.

Proposition 3.10.2. *If H is a group, then $[y, x]_r^{-1} = [x, y]_r = [x^{-1}, y^{-1}]_l = [y^{-1}, x^{-1}]_l^{-1}$, for every x, y in H.*

Proof. It is straightforward. ∎

Example 3.10.3. Suppose that $H = \{e, a, b\}$. Consider the hypergroup $(H, \cdot)$, where $\cdot$ is defined on H as follows:

$\cdot$	e	a	b
e	$\{a, b\}$	e	e
a	e	a	b
b	e	$\{a, b\}$	$\{a, b\}$

It is easy to see that $\{a\} = [a, a]_r \neq [a, a]_l = \{a, b\} = [a^{-1}, a^{-1}]_l$, where a^{-1} is the inverse of a in H.

Proposition 3.10.4. *If H is a commutative hypergroup, then $[x, y]_r = [x, y]_l = [x, y]$, for all $(x, y) \in H^2$.*

Proof. It is straightforward. ∎

Let X be a non-empty subset of a polygroup $\langle P, \cdot, e, ^{-1} \rangle$. Let $\{A_i \mid i \in J\}$ be the family of all subpolygroups of P containing X. Then, $\bigcap_{i \in J} A_i$ is called the *subpolygroup generated* by X. This subpolygroup is denoted by $< X >$ and we have $< X > = \cup \{x_1^{\varepsilon_1} \cdot \ldots \cdot x_k^{\varepsilon_k} \mid x_i \in X, k \in \mathbb{N}, \varepsilon_i \in \{-1, 1\}\}$. If $X = \{x_1, x_2, \ldots, x_n\}$, then the subpolygroup $< X >$ is denoted $< x_1, x_2, \ldots, x_n >$. In a special case $< [P, P]_r >$, $< [P, P]_l >$ and $< [P, P] >$ are shown by P'_r, P'_l and P', respectively.

Proposition 3.10.5. *Let $\langle P, \cdot, e, ^{-1} \rangle$ be a polygroup and $(x, y) \in P^2$. Then,*

(1) $[x, y]_r = [x^{-1}, y^{-1}]_l$;
(2) $P' = P'_r = P'_l$;
(3) $x \in P' \Rightarrow x^{-1} \in P'$.

Proof. (1) Suppose that $u \in [x, y]_r$. Then, $x \cdot y \cap y \cdot x \cdot u \neq \emptyset$ so there exists $t \in P$ such that $t \in x \cdot y \cap y \cdot x \cdot u$. Thus, $t \in x \cdot y \cap v \cdot u$ for some $v \in y \cdot x$. Since P is a polygroup, we have $v^{-1} \in x^{-1} \cdot y^{-1} \cap u \cdot y^{-1} \cdot x^{-1}$. Therefore,

$u \in [x, y]_l$. Hence, $[x, y]_r \subseteq [x, y]_l$. Similarly, we have $[x, y]_l \subseteq [x, y]_r$.

(2) It follows from (1).

(3) Suppose that $a \in [x, y]_r$. Then, $x \cdot y \cap y \cdot x \cdot a \neq \emptyset$ and so $y^{-1} \cdot x^{-1} \cap a^{-1} \cdot x^{-1} \cdot y^{-1} \neq \emptyset$. Hence, $a^{-1} \in [y^{-1}, x^{-1}]_l$. Now, let $x \in P'$. Then, we have $x \in x_1 \cdot x_2 \cdot \ldots \cdot x_n$, where $x_i \in [a_i, b_i]_r$ and $1 \leq i \leq n$. Therefore, by (2) $x^{-1} \in x_n^{-1} \cdot x_{n-1}^{-1} \cdot \ldots \cdot x_1^{-1} \subseteq P'$ follows. ∎

Corollary 3.10.6. *If $\langle P, \cdot, e, ^{-1} \rangle$ is a polygroup, then P' is a subpolygroup of P.*

From now on we call P' the *derived subpolygroup* of P.

Proposition 3.10.7. *Let $\langle P, \cdot, e, ^{-1} \rangle$ be a polygroup. Then, $P' = \{e\}$ if and only if P be an abelian group.*

Proof. Suppose that $P' = \{e\}$. Then, we have $a \cdot b \cdot a^{-1} \cdot b^{-1} = e$, for all $(a, b) \in P^2$ and hence there exists $x \in P$ such that $e = a \cdot x$. Thus, $x = a^{-1}$ and so $a \cdot a^{-1} = e$. Now, let $(x, y) \in P^2$ and $z \in x \cdot y$. We obtain $y \in x^{-1} \cdot z$. Therefore, $x \cdot y = z$ and hence P is an abelian group. ∎

Definition 3.10.8. Let $\langle P, \cdot, e, ^{-1} \rangle$ be a polygroup. Then, we define

$$N(P') = \{u \in P' \mid u \cdot P' = P' \cdot u\}$$

and is called the *normalizer P' in P*.

Example 3.10.9. Suppose that $P = \{e, a, b, c\}$. We consider the non-commutative polygroup $\langle P, \cdot, e, ^{-1} \rangle$, where $\cdot$ is defined on P as follow:

$\cdot$	e	a	b	c
e	e	a	b	c
a	a	a	P	c
b	b	$\{e, a, b\}$	b	$\{b, c\}$
c	c	$\{a, c\}$	c	P

In this case, we can see that $P' = P$.

Recall that a subhypergroup K of a hypergroup $(H, \cdot)$ is invertible on the right if and only if $K \setminus H = \{K \cdot x \mid x \in H\}$ is a partition of H. If K is an invertible to the left subhypergroup of a hypergroup H, then the quotient $K \setminus H$ by the hyperoperation

$$K \cdot x \otimes K \cdot y = \{K \cdot z \mid z \in x \cdot K \cdot y\},$$

is a hypergroup.

Theorem 3.10.10. *If $\langle P, \cdot, e, ^{-1} \rangle$ is a polygroup, then*

(1) P' is invertible;

(2) $N = N(P')$ is subpolygroup of $\langle P, \cdot, e, ^{-1} \rangle$;

(3) $P' \trianglelefteq N(P)$ and $(P' \setminus N, \otimes)$ is commutative polygroup;

(4) if K is a complete subpolygroup of P such that $K \setminus P$ is commutative, then $P' \subseteq K$.

Proof. (1) It is obvious.

(2) Suppose that $x, y \in N$ and $a \in x \cdot y$. We have $a \cdot P' \subseteq x \cdot y \cdot P' = x \cdot P' \cdot y = P' \cdot x \cdot y$. Hence, $a \cdot P' = P' \cdot b$, for some $b \in x \cdot y$. Therefore, $a \in P' \cdot b$ and so $P' \cdot a = P' \cdot b = a \cdot P'$ which means $a \in N$. Moreover, if $x \in N$ we can easily see that $x^{-1} \in N$.

(3) It is easy to see that the hyperoperation $\otimes$ is well defined and also $P' \setminus N$ is a polygroup. For commutativity, since P is a polygroup, we have $[x, y]_t = x \cdot y \cdot x^{-1} \cdot y^{-1}$, for all $(x, y) \in P^2$. Hence, $x \cdot y \subseteq [x, y]_t \cdot y \cdot x$ and so $N \cdot x \otimes N \cdot y = N \cdot x \cdot y = N \cdot y \cdot x = N \cdot y \otimes N \cdot x$.

(4) Suppose that $(x, y) \in P^2$ and $a \in x \cdot y \cdot x^{-1} \cdot y^{-1} = [x, y]_t$. Since $K \subseteq K \cdot x \cdot y \cdot x^{-1} \cdot y^{-1}$, we have $x \cdot y \cdot x^{-1} \cdot y^{-1} \cap K \neq \emptyset$. Hence, $x \cdot y \cdot x^{-1} \cdot y^{-1} \subseteq K$ and so $a \in K$. Therefore, $P' \subseteq K$. ∎

Definition 3.10.11. A polygroup P is called *perfect* if $P' = P$.

Definition 3.10.12. A polygroup P is called *solvable* if $P^{(n)} = \omega_P$, for some $n \in \mathbb{N}$, where $P^{(1)} = P'$ and $P^{(n+1)} = (P^{(n)})'$.

Proposition 3.10.13. *Every non-trivial perfect group is not solvable.*

Proof. It is straightforward. ∎

In the following, we show that the above proposition is not true for the class of polygroups.

Example 3.10.14. Suppose that $P = \{e, a, b, c\}$. Consider the commutative polygroup $\langle P, \cdot, e, ^{-1} \rangle$, where $\cdot$ is defined on P as follows:

$\cdot$	e	a	b	c
e	e	a	b	c
a	a	P	$\{a, b, c\}$	$\{a, b, c\}$
b	b	$\{a, b, c\}$	P	$\{a, b, c\}$
c	c	$\{a, b, c\}$	$\{a, b, c\}$	P

We can easily see that P is a perfect and solvable polygroup. Notice that $P' = P = \omega_P$.

Example 3.10.15. Suppose that $P = \{e, a, b, c\}$. Consider the non-commutative polygroup $\langle P, \cdot, e, ^{-1} \rangle$, where $\cdot$ is defined on P as follows:

$\cdot$	e	a	b	c
e	e	a	b	c
a	a	a	P	c
b	b	$\{e, a, b\}$	b	$\{b, c\}$
c	c	$\{a, c\}$	c	P

In this case, we can see that $P' = P = \omega_P$.

In the following theorem, consider $G//H$ as a double coset algebra (see Example 3.1.2).

Theorem 3.10.16. *Let $(G, \cdot)$ be a group and H be a subgroup of G. We set $HG'H = \{HgH \mid g \in G'\}$. Then,*

(1) $HG'H \subseteq (G//H)'$;
(2) If $G' \cdot H = G$ then $(G//H)$ is a perfect polygroup;
(3) If $HG'H = (G//H)$ then $G' \cdot H = G$.

Proof. (1) We have

$$[HaH, HbH] = (HaH) * (HbH) * (HaH)^{-I} * (HbH)^{-I}$$
$$= \{Hah_1bh_2a^{-1}h_3b^{-1}H \mid h_1, h_2, h_3 \in H\},$$

since H is a subgroup of G, $e \in H$. Now, suppose that $h_1 = h_2 = h_3 = e$. Then, $H[a, b]H = Haba^{-1}b^{-1}H \in [HaH, HbH]$. Therefore, $HG'H \subseteq (G//H)'$.

(2) Suppose that $HxH \in (G//H)$. Then, $x \in G$. Hence, there exists $a, b \in G$ and $h \in H$ such that $x = [a, b]h$. So, we have $HxH = H[a, b]hH = H[a, b]H \in HG'H \subseteq (G//H)'$. Thus, $(G//H)$ is a perfect polygroup.

(3) Since G' is a normal subgroup of G, then $G' \cdot H \subseteq G$. Now, suppose that $x \in G$. Then, there exists $(a, b) \in G^2$ such that $HxH = H[a, b]H$. Therefore, there exists $(h_1, h_2) \in H^2$ such that $x = h_1[a, b]h_2 = h_1[a, b]h_1^{-1}h_1h_2 = [h_1ah_1^{-1}, h_1bh_1^{-1}]h_1h_2 \in G' \cdot H$. Hence, $G = G' \cdot H$. ∎

Definition 3.10.17. Let P be a hypergroup. Suppose that $\tau = \bigcup_{m \geq 1} \tau_m$, where τ_1 is the diagonal relation and for every integer $m > 1$, τ_m is the relation defined as follows:

$$x \, \tau_m \, y \Leftrightarrow \exists (z_1, \ldots, z_m) \in P^m, \exists \sigma \in \mathbb{S}_m : \sigma(i) = i \text{ if } z_i \notin P' \text{ such that}$$
$$x \in \prod_{i=1}^{m} z_i \text{ and } y \in \prod_{i=1}^{m} z_{\sigma(i)}.$$

Obviously, the relation τ is reflexive and symmetric. Now, let τ^* be the transitive closure of τ.

Theorem 3.10.18. *The relation τ^* is a strongly regular relation.*

Proof. It is easy to see that τ^* is an equivalence relation. In order to prove that it is strongly regular, first we show that:

$$x\tau y \Rightarrow x \cdot z \,\overline{\overline{\tau^*}}\, y \cdot z, \quad z \cdot x \,\overline{\overline{\tau^*}}\, z \cdot y,$$

for every $z \in P$. Suppose that $x\tau y$. Then, there exists $m \in \mathbb{N}$ such that $x\tau_m y$. Hence, there exist $(z_1, \ldots, z_m) \in P^m$, $\sigma \in \mathbb{S}_m$ with $\sigma(i) = i$ if $z_i \notin P'$, such that $x \in \prod_{i=1}^{m} z_i$ and $y \in \prod_{i=1}^{m} z_{\sigma(i)}$. Suppose that $z \in P$. We have $x \cdot z \subseteq (\prod_{i=1}^{m} z_i) \cdot z$, $y \cdot z \subseteq (\prod_{i=1}^{m} z_{\sigma(i)}) \cdot z$ and $\sigma(i) = i$ if $z_i \notin P'$. Now, suppose that $z_{i+1} = z$ and we define the permutation $\sigma' \in \mathbb{S}_{m+1}$ as follows:

$$\sigma'(i) = \sigma(i), \text{ for all } 1 \le i \le m \text{ and } \sigma'(m+1) = m+1.$$

Thus, $x \cdot z \subseteq \prod_{i=1}^{m+1} z_i$ and $y \cdot z \subseteq \prod_{i=1}^{m+1} z_{\sigma'(i)}$ such that $\sigma'(i) = i$ if $z_i \notin P'$. Therefore, $x \cdot z \,\overline{\overline{\tau^*}}\, y \cdot z$. Similarly, we have $z \cdot x \,\overline{\overline{\tau^*}}\, z \cdot y$. Now, if $x\tau^* y$ then there exists $k \in \mathbb{N}$ and $(x = u_0, u_1, \ldots, u_k = y) \in P^{k+1}$ such that $x = u_0 \tau u_1 \tau \ldots \tau u_{k-1} \tau u_k = y$. Hence, by the above results, we obtain

$$x \cdot z = u_0 \cdot z \,\overline{\overline{\tau^*}}\, u_1 \cdot z \,\overline{\overline{\tau^*}}\, u_2 \cdot z \,\overline{\overline{\tau^*}}\, \ldots \,\overline{\overline{\tau^*}}\, u_{k-1} \cdot z \,\overline{\overline{\tau^*}}\, u_k \cdot z = y \cdot z$$

and so $x \cdot z \,\overline{\overline{\tau^*}}\, y \cdot z$.

Similarly, we can prove that $z \cdot x \,\overline{\overline{\tau^*}}\, z \cdot y$. Therefore, τ^* is a strongly regular relation on P. ■

Definition 3.10.19. Let $(H, \cdot)$ and $(K, *)$ be hypergroups. A map $f : H \to P^*(K)$ is called a *good multihomomorphism* if

$$f(x \cdot y) = f(x) * f(y),$$

for all $x, y \in H$. If $(H, \cdot) = (K, *)$ and $\bigcup_{h \in H} f(h) = H$, then f is called a *generalized automorphism*. Moreover, we will denote by $GAut(H)$ the set of all generalized automorphisms of $(H, \cdot)$.

Proposition 3.10.20. $(GAut(H), \circ)$ *is a monoid, where $\circ$ is defined as follows:*

$$(f \circ g)(h) = \bigcup_{a \in g(h)} f(a),$$

for all $f, g \in GAut(H)$ and $h \in H$. Moreover, $Aut(H)$ (i.e., the group of automorphism of H) is a subgroup of $GAut(H)$.

Proof. It is straightforward. ∎

Definition 3.10.21. Let $(H, \cdot)$ and $(K, *)$ be two hypergroups. We consider the monoid $GAut(H)$ and the group $\frac{K}{\tau^*}$. Let:

$$\varphi : \; \frac{K}{\tau^*} \to GAut(H),$$

$$\tau^*(x) \to \varphi_{\tau^*(x)}$$

be a homomorphism. Then, we define a hyperoperation in $H \times K$ as follows:

$$(x_1, y_1) \circ (x_2, y_2) = \{(x, y) \mid x \in x_1 \cdot \varphi_{\tau^*(y_1)}(x_2), y \in y_1 * y_2\}.$$

We call it a τ-*multisemi-direct hyperproduct* of hypergroups H and K through φ and we denote it by $H \times_\varphi K$. Moreover, we call a τ-multisemi-direct hyperproduct is special if $Im(\varphi) \subseteq Aut(H)$.

Lemma 3.10.22. *Let H and K be two hypergroups. Then, $H \times K$ equipped with the τ-multisemi-hyperproduct is a hypergroup.*

Proof. Suppose that (x_1, y_1), (x_2, y_2) and (x_3, y_3) are elements in $H \times K$. If $(s, t) \in [(x_1, y_1) \circ (x_2, y_2)] \circ (x_3, y_3)$, then $(s, t) \in (u, v) \circ (x_3, y_3)$ for some $(u, v) \in (x_1, y_1) \circ (x_2, y_2)$. Therefore, $s \in u \cdot \varphi_{\tau^*(v)}(x_3), t \in v * y_3$ and $u \in x_1 \cdot \varphi_{\tau^*(y_1)}(x_2), v \in y_1 * y_2$. Thus, $s \in (x_1 \cdot \varphi_{\tau^*(y_1)}(x_2)) \cdot \varphi_{\tau^*(v)}(x_3)$ and $t \in (y_1 * y_2) * y_3$. By the associativity of $\cdot$ and $*$, we have $s \in x_1 \cdot (\varphi_{\tau^*(y_1)}(x_2) \cdot \varphi_{\tau^*(v)}(x_3))$ and $t \in y_1 * (y_2 * y_3)$. Since $v \in y_1 * y_2$, we conclude that $s \in x_1 \cdot (\varphi_{\tau^*(y_1)}(x_2) \cdot \varphi_{\tau^*(y_1)\tau^*(y_2)}(x_3))$. On the other hand, if $(s, t) \in (x_1, y_1) \circ [(x_2, y_2) \circ (x_3, y_3)]$, then $(s', t') \in (x_1, y_1) \circ (u', v')$ for some $(u', v') \in (x_2, y_2) \circ (x_3, y_3)$. Therefore, $s' \in x_1 \cdot \varphi_{\tau^*(y_1)}(u'), t' \in y_1 * v'$ and $u' \in x_2 \cdot \varphi_{\tau^*(y_2)}(x_3), v' \in y_2 * y_3$. Thus,

$$s' \in x_1 \cdot \varphi_{\tau^*(y_1)}(x_2 \cdot \varphi_{\tau^*(y_2)}(x_3)) = x_1 \cdot (\varphi_{\tau^*(y_1)}(x_2) \cdot \varphi_{\tau^*(y_1)\tau^*(y_2)}(x_3))$$

and $t' \in y_1 * (y_2 * y_3)$. By the above results, we conclude the associativity of $\circ$.

Now, suppose that $(a, x) \in H^2$ and $(b, y) \in K^2$. Since H and K are hypergroups, there exists $(u, w) \in H \times K$ such that $a \in x \cdot u$ and $b \in y * w$. Since $u \in H = \varphi_{\tau^*(y)}(K)$, we conclude that there exists $t \in K$ such that $u \in \varphi_{\tau^*(y)}(t)$ and so we have $(a, b) \in (x, y) \circ (t, w)$. Similarly, there exists $(t', w') \in H \times K$ such that $(a, b) \in (t', w') \circ (x, y)$. ∎

Theorem 3.10.23. *Let P_1 and P_2 be two polygroups. Then, $P_1 \times P_2$*

equipped with a special τ-multisemi-hyperproduct is a ploygroup.

Proof. Suppose that φ is a special τ-multisemi-hyperproduct. According to the previous lemma, $P_1 \times_\varphi P_2$ is a hypergroup. Since φ is special, $a = a \cdot \varphi_{\tau^*(b)}(e_1)$ and $a = e_1 \cdot \varphi_{\tau^*(e_1)}(b)$. Hence, $(a, b) \circ (e_1, e_2) = (a, b) = (e_1, e_2) \circ (a, b)$, that is, (e_1, e_2) is the identity element in $P_1 \times_\varphi P_2$. Moreover, we can check that $(\varphi_{\tau^*(b^{-1})}(a^{-1}), b^{-1})$ is the inverse element of $(a, b) \in P_1 \times P_2$. Now, we check that if $(z_1, z_2) \in (x_1, x_2) \circ (y_1, y_2)$, then $(x_1, x_2) \in (z_1, z_2) \circ (y_1, y_2)^{-1}$. Since $z_2 \in x_2 * y_2$ and $z_1 \in x_1 \cdot \varphi_{\tau^*(x_2)}(y_1)$, we have

$$(z_1, z_2) \circ (y_1, y_2)^{-1} = \{(a, b) | a \in z_1 \cdot \varphi_{\tau^*(z_2)}(\varphi_{\tau^*(y_2^{-1})}(y_1^{-1})), b \in z_2 * y_2^{-1}\}$$

$$= \{(a, b) | a \in z_1 \cdot \varphi_{\tau^*(z_2)\tau^*(y_2^{-1})}(y_1^{-1}), b \in z_2 * y_2^{-1}\}$$

$$= \{(a, b) | a \in z_1 \cdot \varphi_{\tau^*(x_2)}(y_1^{-1}), b \in z_2 * y_2^{-1}\}.$$

Hence, $x_1 \in z_1 \cdot \varphi_{\tau^*(x_2)}(y_1)^{-1} = z_1 \cdot \varphi_{\tau^*(x_2)}(y_1^{-1})$ and $x_2 \in z_2 * y_2^{-1}$ as we want. ∎

The τ-multisemi-direct hyperproduct of polygroups P_1 and P_2 through zero homomorphism φ_0, i.e.,

$$\varphi_0 : \frac{P_2}{\tau^*} \to GAut(P_1),$$

$$\varphi_0(\tau^*(x)) = i_{Aut(P_1)}$$

which we denote it by $P_1 \times P_2$ and is called τ-*direct hyperproduct* of P_1 and P_2.

Proposition 3.10.24. *Let P_1, P_2 be two polygroups. Then, $(P_1 \times P_2)' = P_1' \times P_2'$.*

Proof. Suppose that $((x_1, y_1), (x_2, y_2)) \in (P_1 \times P_2)^2$. Then, we have

$$[(x_1, y_1), (x_2, y_2)] = (x_1, y_1) \otimes (x_2, y_2) \otimes (x_1^{-1}, y_1^{-1}) \otimes (x_2^{-1}, y_2^{-1})$$

$$= \{(x, y) | x \in x_1 \cdot x_2 \cdot x_1^{-1} \cdot x_2^{-1}, y \in y_1 \circ y_2 \circ y_1^{-1} \circ y_2^{-1}\}$$

$$= \{(x, y) | x \in [x_1, x_2], y \in [y_1, y_2]\}$$

$$= [x_1, x_2] \times [y_1, y_2],$$

and the proof completes. ∎

Corollary 3.10.25. τ-*direct hyperproduct of two polygroups P_1 and P_2 is perfect if and only if P_1 and P_2 are perfect.*

Let H be a regular hypergroup. For $n \in \mathbb{N}$, let $a_1, \ldots, a_n$ be elements in H, and $a_1', \ldots, a_n'$ are their inverses in H, respectively. The set

$$a_1 \cdot a_2 \cdot \ldots \cdot a_n \cdot a_n' \cdot a_{n-1}' \cdot \ldots \cdot a_1'$$

is called a *product of type zero* and denote with $N(0)$ the union of all products of type 0.

Theorem 3.10.26. *Let H be a regular and reversible hypergroup and e be a bilateral identity. If $e \in \prod_{i=1}^{n} z_i$, then there exist inverses of $z_1, \ldots, z_n$ respectively $z_1', \ldots, z_n'$ such that $z_n \in z_{n-1}' \cdot z_{n-2}' \cdot \ldots \cdot z_1' \cdot e$.*

Proof. The theorem is true for $n = 2$. Indeed, for the hypothesis of reversibility, $e \in z_1 \cdot z_2$ implies that there exists at least one inverse of z_1, like z_1' such that $z_2 \in z_1' \cdot e$. Suppose that for for every $k < n$ and for every $y_1, \ldots, y_k \in H$, the implication of the theorem is satisfied that is there exist $y_1', \ldots, y_{k-1}'$ such that $y_k \in y_{k-1}' \cdot \ldots \cdot y_1' \cdot e$. The hypothesis $e \in \prod_{i=1}^{n} z_i$ implies that there exists $u \in z_{n-1} \cdot z_n$ such that $e \in \prod_{i=1}^{n-1} z_i \cdot u$. From $u \in z_{n-1} \cdot z_n$ follow that there exists z_{n-1}' such that $z_n \in z_{n-1}' \cdot u$. Since $e \in \prod_{i=1}^{n-2} z_i \cdot u$, the inductive hypothesis implies that there exist $z_1', \ldots, z_{n-2}'$ such that $u \in z_{n-2}' \cdot \ldots \cdot z_1' \cdot e$. Therefore, $z_n \in z_{n-1}' \cdot z_{n-2}' \cdot \ldots \cdot z_1' \cdot e$. ∎

Theorem 3.10.27. *If H is a regular and reversible hypergroup, then the heart of H is the union of the products of type zero (i.e., $\omega_H = N(0)$).*

Proof. It is clear that $N(0) \subseteq \omega_H$. We prove the converse. Let $x \in \omega_H$. Then, if e is a bilateral identity, one has $x\beta^*e$. Now, it follows that there exists $n \geq 1$ and $z_1, \ldots, z_n$ such that $\{e, x\} \subseteq \prod_{i=1}^{n} z_i$. By Theorem 3.10.26, there exist $z_1', \ldots, z_{n-1}'$ such that $x \in z_1 \cdot \ldots \cdot z_{n-1} \cdot z_{n-1}' \cdot \ldots z_1' \cdot e \subseteq e \cdot z_1 \cdot \ldots \cdot z_{n-1} \cdot z_{n-1}' \cdot \ldots \cdot z_1' \cdot e \subseteq N(0)$. ∎

Corollary 3.10.28. *Let ω_{P_1}, ω_{P_2} and $\omega_{P_1 \times P_2}$ be the hearts of P_1, P_2 and $P_1 \times P_2$, respectively. Then, $\omega_{P_1 \times P_2} = \omega_{P_1} \times \omega_{P_2}$.*

Proposition 3.10.29. *Let P_1, P_2 be two polygroups. If $P_1 \times P_2$ is solvable, then P_1 and P_2 are solvable.*

Proof. The proof follows from Proposition 3.10.24 and Corollary 3.10.28. ∎

3.11 Nilpotent polygroups

Jafarpour, Aghabozorgi and Davvaz in [84] introduced the concept of nilpotent polygroups. In this section, we study the notions of nilpotent polygroups by using the notion of hearts of polygroups. In particular, we obtain a necessary and sufficient condition between nilpotent (solvable) polygroups and fundamental groups.

Let $(G, \cdot)$ be a group and $P_G = G \cup \{a\}$, where $a \notin G$. We define on P_G the hyperoperations $\circ$ as follows:

(1) $a \circ a = e$;
(2) $e \circ x = x \circ e = x$, for every $x \in P_G$;
(3) $a \circ x = x \circ a = x$, for every $x \in P_G \setminus \{e, a\}$;
(4) $x \circ y = x \cdot y$, for every $(x, y) \in G^2$ such that $y \neq x^{-1}$;
(5) $x \circ x^{-1} = \{e, a\}$, for every $x \in P_G \setminus \{e, a\}$.

Proposition 3.11.1. *If G is a group, then $\langle P_G, \circ, e, ^{-1} \rangle$ is a polygroup.*

Proof. First of all, we prove the associativity of $\circ$. Suppose that $(x, y, z) \in P_G^3$.

(i) If $\{x, y, z\} \cap \{e, a\} = \emptyset$, then we have two following cases.
    Case1. $x \neq y^{-1} \neq z$ and $x \neq z^{-1}$. In this case $(x \circ y) \circ z = (x \cdot y) \cdot z = x \cdot (y \cdot z) = x \circ (y \circ z)$.
    Case2. There exists $\{u, v\} \subseteq \{x, y, z\}$ such that $u = v^{-1}$. Without losing generality suppose that $x = u, y = v$. Thus, $(x \circ y) \circ z = \{e, a\} \circ z$. Hence, $\{e, a\} \circ z = z$. On the other hand, if $y = z^{-1}$, then $x \circ (y \circ z) = x \circ \{e, a\} = x = y^{-1} = z$ and if $y \neq z^{-1}$ we have $x \circ (y \circ z) = x \circ (y \cdot z) = x \cdot (y \cdot z) = (x \cdot x^{-1}) \cdot z = e \cdot z = z$.
(ii) If $\{x, y, z\} \cap \{e, a\} \neq \emptyset$. Let $e \in \{x, y, z\}$. It is easy to see that the associativity condition holds. Now suppose that $\{x, y, z\} \cap \{e, a\} = \{a\}$. Without losing generality let $x = a$, in this case we have

$$(x \circ y) \circ z = x \circ (y \circ z) = \begin{cases} a & \text{if } y = a, z = a \\ z & \text{if } y = a, z \neq a \\ y & \text{if } y \neq a, z = a \\ y \cdot z & \text{if } y \neq z^{-1}, y \neq a \neq z \\ \{e, a\} & \text{if } y = z^{-1}, y \neq a \neq z. \end{cases}$$

According to the structure of $\circ$ we conclude that e is the identity element of P_G and the other conditions for being polygroup hold too. ∎

Proposition 3.11.2. *If G is a group, then $P_G / \beta^* \cong G$.*

Proof. It is straightforward. ∎

Definition 3.11.3. A polygroup $\langle P, \cdot, e, ^{-1} \rangle$ is said to be *nilpotent* if

$\ell_n(P) \subseteq \omega_P$ or equivalently $\ell_n(P) \cdot \omega_P = \omega_P$, for some integer n, where $\ell_0(P) = P$ and

$$\ell_{k+1}(P) = \langle \{h \in P \mid x \cdot y \cap h \cdot y \cdot x \neq \emptyset, \text{ such that } x \in \ell_k(P) \text{ and } y \in P\} \rangle.$$

The smallest integer c such that $\ell_c(P) \cdot \omega_P = \omega_P$ is called the *nilpotency class* or for simplicity the *class* of P.

Notice that $P = \ell_0(P) \supseteq \ell_1(P) \supseteq \ell_2(P) \supseteq \ldots$ that is $\{\ell_k(P)\}_{k \geq o}$ is a decreasing sequence which we call it *generalized descending central series*.

Proposition 3.11.4. *Every commutative polygroup is nilpotent of class* 1.

Proof. Suppose that $\langle P, \cdot, e, ^{-1} \rangle$ is a commutative polygroup and $h \in \ell_1(P)$. Then, there exists $(x, y) \in P^2$ such that $x \cdot y \cap h \cdot y \cdot x \neq \emptyset$. Since P is commutative we have $x \cdot y \cap h \cdot x \cdot y \neq \emptyset$. So, $\bar{x}\bar{y} = \bar{h}\bar{x}\bar{y}$ and so $\bar{h} = e$ which means that $h \in \omega_P$. Therefore, $\ell_1(P) \cdot \omega_P = \omega_P$ ∎

A polygroup is called *proper* if it not a group.

Proposition 3.11.5. *Every proper polygroup of order less than* 7 *is nilpotent of class* 1.

Proof. Suppose that $\langle P, \cdot, e, ^{-1} \rangle$ is a polygroup of order less than 7. Then, P/β^* is an abelian group of order less that 6. Now, let $h \in \ell_1(P)$. Then, there exists $(x, y) \in P^2$ such that $x \cdot y \cap h \cdot x \cdot y \neq \emptyset$. Thus, $\bar{x}\bar{y} = \bar{h}\bar{y}\bar{x} = \bar{h}\bar{x}\bar{y}$ which implies that $h \in \omega_P$. Therefore, $\ell_1(P) \subseteq \omega_P$ and consequently $\ell_1(P) \cdot \omega_P = \omega_P$. ∎

Corollary 3.11.6. *The symmetric group* S_3 *is the smallest non-nilpotent polygroup.*

Example 3.11.7. Let $P = \{e, a, b, c, d, f, g\}$. We consider the proper non-commutative polygroup $\langle P, \cdot, e, ^{-1} \rangle$, where $\cdot$ is defined on P as follows:

$\cdot$	e	a	b	c	d	f	g
e	e	a	b	c	d	f	g
a	a	e	b	c	d	f	g
b	b	b	$\{e,a\}$	g	f	d	c
c	c	c	f	$\{e,a\}$	g	b	d
d	d	d	g	f	$\{e,a\}$	c	b
f	f	f	c	d	b	g	$\{e,a\}$
g	g	g	d	b	c	$\{e,a\}$	f

It is easy to see that $\omega_P = \{e,a\}$ while $\ell_n(P) = \{e,a,f,g\}$ and hence $\ell_n(P) \cdot \omega_P \neq \omega_P$ for all $n \in \mathbb{N}$. Thus, P is not a nilpotent polygroup of order 7.

Proposition 3.11.8. *Let $\langle P, \cdot, e, ^{-1} \rangle$ be a polygroup and $G = \frac{P}{\beta^*}$. Then, for all $k \geq 1$*

$$\ell_k(G) = \langle \bar{t} \mid t \in \ell_k(P) \rangle.$$

Proof. Suppose that $\langle P, \cdot, e, ^{-1} \rangle$ is a polygroup and $G = \frac{P}{\beta^*}$. Then, we do the proof by induction on k. For $k = 0$, we have $\langle \bar{t} \mid t \in \ell_0(P) = P \rangle = \ell_0(G)$. Now, suppose that $\bar{a} \in \langle \bar{t} \mid t \in \ell_{k+1}(P) \rangle$. Then, $a \in \ell_{k+1}(P)$ and so there exist $x \in \ell_k(P)$ and $y \in P$ such that $xy \cap ayx \neq \emptyset$. Thus, $\bar{x}\bar{y} = \bar{a}\bar{y}\bar{x}$. By hypothesis of induction we conclude that $\bar{a} \in \ell_{k+1}(G)$. Conversely, let $\bar{a} \in \ell_{k+1}(G)$. Without losing generality suppose that $\bar{a} = \bar{x}\bar{y}\bar{x}^{-1}\bar{y}^{-1}$, where $\bar{x} \in \ell_k(G), \bar{y} \in G$, which implies that $\bar{x}\bar{y} = \bar{a}\bar{y}\bar{x}$. Thus, there exist $c \in xy$ and $d \in ayx$ such that $\bar{c} = \bar{d}$. Since P is a polygroup there exists $u \in P$ such that $c \in xy \cap uyx$. From $\bar{x} \in \ell_k(G), \bar{y} \in G$ and the hypothesis of induction we have $x \in \ell_k(P), y \in P$. Thus, $u \in \ell_{k+1}(P)$ and $\bar{a}\bar{y}\bar{x} = \bar{d} = \bar{c} = \bar{x}\bar{y} = \bar{u}\bar{y}\bar{x}$. Therefore, $\bar{a} = \bar{u} \in \langle \bar{t} \mid t \in \ell_{k+1}(P) \rangle$. ∎

Theorem 3.11.9. *Let $\langle P, \cdot, e, ^{-1} \rangle$ be a polygroup. Then, P is nilpotent if and only if $G = \frac{P}{\beta^*}$ is nilpotent.*

Proof. Suppose that P is a nilpotent polygroup so there exists $k \in \mathbb{N}$ such that $\ell_k(P) \subseteq \omega_p$. According to the previous proposition, we have $\ell_k(G) = \langle \bar{t} \mid t \in \ell_k(P) \subseteq \omega_p \rangle = \{e_G\} = \omega_G$, and so $G = \frac{P}{\beta^*}$ is a nilpotent group. Similarly, we can see the converse. ∎

Corollary 3.11.10. *Let G be a group. Then, P_G is nilpotent if and only if G is nilpotent.*

Theorem 3.11.11. *Let $\langle P, \cdot, e, ^{-1} \rangle$ be a polygroup and N be a normal subpolygroup of P. Then, $\ell_n(\frac{P}{N}) = \frac{\ell_n(P) \cdot N}{N}$, for all $n \geq 0$.*

Proof. By induction on n we show that $\ell_n(\frac{P}{N}) \subseteq \frac{\ell_n(P) \cdot N}{N}$ and $\frac{\ell_n(P) \cdot N}{N} \subseteq \ell_n(\frac{P}{N})$. For $n = 0$, the inclusions are obvious. Now, suppose that $yN \in \ell_{n+1}(\frac{P}{N})$. Then, $yN \in [aN, bN]$, where $aN \in \ell_n(\frac{P}{N})$ and $bN \in \frac{P}{N}$. By the hypothesis of induction we have $aN = a'N$, where $a' \in \ell_n(P)$. Therefore, $yN = y'N$, where $y' \in [a', b]$. Thus, $yN \in \frac{\ell_{n+1}(P) \cdot N}{N}$. If $yN \in \frac{\ell_{n+1}(P) \cdot N}{N}$, then $yN = y''N$, where $y'' \in \ell_{n+1}(P)$. So, there exist $a \in \ell_n(P)$ and $b \in P$ such that $y'' \in [a, b]$. Hence, $aN = yN \in [aN, bN]$, where $aN \in \ell_n(\frac{P}{N})$ which means that $yN \in \ell_{n+1}(\frac{P}{N})$ and our proof is completed. ∎

Corollary 3.11.12. *If $\langle P, \cdot, e, ^{-1} \rangle$ is a nilpotent polygroup, then*

(1) every subpolygroup of P is nilpotent;

(2) if N is a normal subpolygroup of P, then $\frac{P}{N}$ is nilpotent.

Definition 3.11.13. Let $\langle P, \cdot, e, ^{-1} \rangle$ be a polygroup. We define $Z_0(P) = \omega_P$ and $Z_n(P) = \langle \{x | x \cdot y \cdot Z_{n-1}(P) = y \cdot x \cdot Z_{n-1}(P), \forall y \in P\} \rangle$, for all $n \in \mathbb{N}$.

Notice that $\omega_P = Z_0(P) \subseteq Z_1(P) \subseteq Z_2(P) \subseteq \ldots$ that is $\{Z_m(P)\}_{m \geq 0}$ is an increasing sequence which we call it *generalized ascending central series*. Moreover, $Z_n(P)$ is a subpolygroup of P, for every $n \geq 0$.

Proposition 3.11.14. *If $\langle P, \cdot, e, ^{-1} \rangle$ is a polygroup and $n \geq 0$, then*

(1) $Z_n(P)$ is a complete subpolygroup of P;

(2) $g \cdot g^{-1} \subseteq Z_n(P)$, for every $g \in P$;

(3) $Z_n(P)$ is a normal subpolygroup of P.

Proof. (1) Since $\omega_P \subseteq Z_n(P)$, we conclude that $C(Z_n(P)) = Z_n(P) \cdot \omega_P = Z_n(P)$, which means that $Z_n(P)$ is complete.

(2) Let $g \in P$. Since $e \in g \cdot g^{-1} \cap Z_n(P)$ and $Z_n(P)$ is complete, $g \cdot g^{-1} \subseteq Z_n(P)$.

(3) Let $g \in P$ be an arbitrary element and $x \in Z_n(P)$. Then, $g \cdot x \cdot g^{-1} \cdot Z_{n-1}(P) = g \cdot g^{-1} \cdot x \cdot Z_{n-1}(P) \subseteq g \cdot g^{-1} \cdot Z_n(P) = Z_n(P)$. Hence, $g \cdot x \cdot g^{-1} \subseteq Z_n(P)$. ∎

Theorem 3.11.15. *Let $\langle P, \cdot, e, ^{-1} \rangle$ be a polygroup. Then, P is nilpotent if and only if there exists $r \geq 0$ such that $Z_r(P) = P$.*

Proof. Suppose that there exists $r \geq 0$ such that $Z_r(P) = P$. In order to prove $\ell_r(P) \subseteq \omega_P$, by induction we show that $\ell_i(P) \subseteq Z_{r-i}(P)$. For $i = 0$ we have $\ell_0(P) = P \subseteq P = Z_r(P)$. Now, if $a \in \ell_{i+1}(P)$, then without loss generality suppose that $x \cdot y \cap a \cdot y \cdot x \neq \emptyset$, where $x \in \ell_i(P)$ and $y \in P$. By the hypothesis of induction we conclude that $x \in Z_{r-i}(P)$. Hence, $x \cdot y \cdot Z_{r-i-1}(P) = y \cdot x \cdot Z_{r-i-1}(P)$ and so $a \in Z_{r-i-1}(P)$. Now if $i = r$, then $\ell_r(P) \subseteq Z_0(P) = \omega_P$. For the converse, suppose that $\ell_r(P) \subseteq \omega_P$. It is enough to show that $\ell_{r-i}(P) \subseteq Z_i(P)$, for all $0 \leq i \leq n$. For $i = 0$, we have $\ell_r(P) \subseteq \omega_P = Z_0(P)$. Let $a \in \ell_{r-i-1}(P)$ and $b \in P$. Then, $[a, b] \subseteq \ell_{r-i}(P)$. By using the hypothesis of induction, we have $[a, b] \subseteq Z_i(P)$. Therefore, $a \cdot b \cdot Z_i(P) = b \cdot a \cdot Z_i(P)$ and so $a \in Z_{i+1}(P)$ as we need. If we take $i = r$, then we get our claim. ∎

Corollary 3.11.16. *Let $\langle P, \cdot, e, ^{-1} \rangle$ be a polygroup. Then, $\ell_c(P) \subseteq \omega_P$ if and only if $Z_c(P) = P$, that is P is nilpotent of class c if and only if*

$Z_c(P) = P$.

Proposition 3.11.17. *Let P_1 and P_2 be two polygroups. Then, for all $k \geqslant 0$*

$$\ell_k(P_1 \times P_2) = \ell_k(P_1) \times \ell_k(P_2).$$

Proof. We prove our claim by induction on k. For $k = 0$, it is obvious. Now, suppose that $(a, b) \in \ell_{k+1}(P_1 \times P_2)$. Then, there exist $(u, v) \in \ell_k(P_1 \times P_2)$ and $(s, t) \in P_1 \times P_2$ such that $(u, v) * (s, t) \cap (a, b) * (s, t) * (u, v) \neq \emptyset$ that is $u \cdot s \cap a \cdot s \cdot u \neq \emptyset$ and $v \circ t \cap b \circ t \circ v \neq \emptyset$. By using the hypothesis of induction, we conclude that $a \in \ell_{k+1}(P_1)$ and $b \in \ell_{k+1}(P_2)$. Thus, $(a, b) \in \ell_{k+1}(P_1) \times \ell_{k+1}(P_2)$. Similarly, we obtain the converse. ∎

Proposition 3.11.18. *Let P_1, P_2 be two polygroups. Then, $P_1 \times P_2$ is nilpotent if and only if P_1 and P_2 are nilpotent.*

Proof. If P_1, P_2 are nilpotent, then there exist k_1 and k_2 such that $\ell_{k_1}(P_1) \subseteq \omega_{P_1}$ and $\ell_{k_2}(P_2) \subseteq \omega_{P_2}$. Suppose that $k = lcm(k_1, k_2)$. Hence, $\ell_k(P_1) \subseteq \ell_{k_1}(P_1)$ and $\ell_k(P_2) \subseteq \ell_{k_2}(P_2)$ and so we obtain $\ell_k(P_1 \times P_2) = \ell_k(P_1) \times \ell_k(P_2) \subseteq \ell_{k_1}(P_1) \times \ell_{k_2}(P_2) \subseteq \omega_{P_1} \times \omega_{P_2} = \omega_{P_1 \times P_2}$. Conversely, suppose that $P_1 \times P_2$ is nilpotent. Then, there exists k such that $\ell_k(P_1) \times \ell_k(P_2) = \ell_k(P_1 \times P_2) \subseteq \omega_{P_1 \times P_2} = \omega_{P_1} \times \omega_{P_2}$. Hence, $\ell_k(P_1) \subseteq \omega_{P_1}$ and $\ell_k(P_2) \subseteq \omega_{P_2}$. Therefore, P_1, P_2 are nilpotent. ∎

Example 3.11.19. Let $P_1 = \{e, a, b, c\}$ be the polygroup in Example 3.10.9 and $P_2 = \{0, 1\}$ be the cyclic group of order two. Consider the non-commutative polygroup $P \cong P_1 \times P_2$, where $\cdot$ is defined on P as follows:

$\cdot$	e	a	b	c	d	f	g	h
e	e	a	b	c	d	f	g	h
a	a	e	c	b	f	d	h	g
b	b	c	b	c	$\{e,b,d,g\}$	$\{a,c,f,h\}$	g	h
c	c	b	c	b	$\{a,c,f,h\}$	$\{e,b,d,g\}$	h	g
d	d	f	$\{e,b,d\}$	a,c,f	d	f	$\{d,g\}$	$\{f,h\}$
f	f	d	$\{a,c,f\}$	$\{e,b,d\}$	f	d	$\{f,h\}$	$\{d,g\}$
g	g	h	$\{b,g\}$	$\{c,h\}$	g	h	$\{e,b,d,g\}$	$\{a,c,f,h\}$
h	h	g	$\{c,h\}$	$\{b,g\}$	h	g	$\{a,c,f,h\}$	$\{e,b,d,g\}$

It is easy to see that $\ell_1(P_1) = P_1' = \omega_{P_1}$ and $\ell_1(P_2) = \omega_{P_2} = \{0\}$. Hence, P is a nilpotent polygroup of class 1.

Proposition 3.11.20. *Let $\langle P_1, \cdot, e_1, {}^{-1} \rangle$ and $\langle P_2, \circ, e_2, {}^{-I} \rangle$ be two polygroups, and $\phi : P_1 \longrightarrow P_2$ be a good homomorphism. If ϕ is one to one* and K_1 is a nilpotent subpolygroup of P_1, then $\phi(K_1)$ is a nilpotent sub-polygroup of P_2.

Proof. (1) By induction on n we show that $\ell_n(\phi(K_1)) = \phi(\ell_n(K_1))$. For $n = 0$ it is obvious. Let $z \in \ell_{n+1}(\phi(K_1))$. Then, there exist $x \in \ell_n(\phi(K_1))$ and $y \in \phi(K_1)$ such that $x \circ y \cap z \circ y \circ x \neq \emptyset$. Since $\{x, y, z\} \subseteq \phi(K_1)$, there exist $a \in \ell_n(\phi(K_1)), b \in K_1, c \in K_1$ such that $\phi(a) = x, \phi(b) = y, \phi(c) = z$. Therefore, we have

$$x \circ y \cap z \circ y \circ x = \phi(a) \circ \phi(b) \cap \phi(c) \circ \phi(b) \circ \phi(a) = \phi(a \cdot b) \cap \phi(c \cdot b \cdot a) \neq \emptyset.$$

Since ϕ is one to one, we conclude that $\phi(a \cdot b \cap c \cdot b \cdot a) \neq \emptyset$, so $a \cdot b \cap c \cdot b \cdot a \neq \emptyset$. By the hypothesis of induction we have $c \in \ell_{n+1}(K_1)$. Thus, $z = \phi(c) \in \phi(\ell_{n+1}(K_1))$.

Conversely, let $z \in \phi(\ell_{n+1}(K_1))$. Then, there exists $c \in \ell_{n+1}(K_1)$ such that $\phi(c) = z$. So, from $c \in \ell_{n+1}(K_1)$ we conclude that there exist $a \in \ell_n(K_1)$ and $b \in K_1$ such that $a \cdot b \cap c \cdot b \cdot a \neq \emptyset$. Thus, $\phi(a) \circ \phi(b) \cap \phi(c) \circ \phi(b) \circ \phi(a) \neq \emptyset$. By the hypothesis of induction we have $\phi(a) \in \phi(\ell_n(K_1) = \ell_n(\phi(K_1))$ and $\phi(b) \in \phi(K_1)$. So, $z = \phi(c) \in \ell_{n+1}(\phi(K_1))$.

Now, let K_1 be a nilpotent subpolygroup of P_1. Then, there exists m such that $\ell_m(K_1) \subseteq \omega_{K_1}$. By the previous theorem, we have $\ell_m(\phi(K_1)) = \phi(\ell_m(K_1) \subseteq \phi(\omega_{K_1}) \subseteq \omega_{\phi(K_1)}$, and the proof is completed. ∎

Proposition 3.11.21. *Every commutative polygroup is solvable of length 1.*

Proof. It is straightforward. ∎

Proposition 3.11.22. *Let P be a polygroup and $G = \frac{P}{\beta^*}$. Then, for all $k \geq 1$*

$$\imath_k(G) = \langle \bar{t} \mid t \in \imath_k(P) \rangle.$$

Proof. Suppose that P is a polygroup and $G = \frac{P}{\beta^*}$. Then, we do the proof by induction on k. For $k = 0$, we have $\langle \bar{t} \mid t \in \imath_0(P) = P \rangle = \imath_0(G)$. Now, suppose that $\bar{a} \in \langle \bar{t} \mid t \in \imath_{k+1}(P) \rangle$. Then, $a \in \imath_{k+1}(P)$ and so there exist $x, y \in \imath_k(P)$ such that $xy \cap ayx \neq \emptyset$. Thus, $\bar{x}\bar{y} = \bar{a}\bar{y}\bar{x}$. By the hypothesis of induction we conclude that $\bar{a} \in \imath_{k+1}(G)$. Conversely, let $\bar{a} \in \imath_{k+1}(G)$. Without losing generality suppose that $\bar{a} = \bar{x}\bar{y}\bar{x}^{-1}\bar{y}^{-1}$, where $\bar{x}, \bar{y} \in \imath_k(G)$, which implies that $\bar{x}\bar{y} = \bar{a}\bar{y}\bar{x}$. Thus, there exist $c \in xy$ and $d \in ayx$ such that $\bar{c} = \bar{d}$. Since P is a polygroup, there exists $u \in P$ such that $c \in xy \cap uyx$. By the hypothesis of induction we have $x, y \in \imath_k(P)$ which implies that $u \in \imath_{k+1}(P)$ and $\bar{a}\bar{y}\bar{x} = \bar{d} = \bar{c} = \bar{x}\bar{y} = \bar{u}\bar{y}\bar{x}$, so $\bar{a} = \bar{u} \in \langle \bar{t} \mid t \in \imath_{k+1}(P) \rangle$. ∎

Theorem 3.11.23. *Let P be a polygroup. Then, P is solvable if and only if $G = \frac{P}{\beta^*}$ is solvable.*

Proof. Suppose that P is a solvable polygroup. Then, there exists $k \in \mathbb{N}$ such that $\imath_k(P) \subseteq \omega_P$. According to the previous proposition, we have $\imath_k(G) = \langle \bar{t} \mid t \in \imath_k(P) \rangle = \{e_G\} = \omega_G$, and so $G = \frac{P}{\beta^*}$ is a nilpotent group. Similarly, we can see the converse. ∎

Corollary 3.11.24. *Every nilpotent polygroup is solvable.*

Proposition 3.11.25. *Every proper polygroup of order less than 61 is solvable.*

Proof. Suppose that $\langle P, \cdot, e, ^{-1} \rangle$ is a proper polygroup of order less than 61. Then, P/β^* is a group of order less that 60. Thus, P/β^* is not isomorphic to the smallest non-solvable group A_5 (alternating group of degree 5). Hence, P is solvable. ∎

In the following example, we introduce one of the smallest proper polygroups of order 61.

Example Let A_5 be the alternating group of degree 5 and $P = A_5 \cup \{a\}$, where $a \notin A_5$. We define on P the hyperoperations $\circ$, as follows:

(1) $a \circ a = \{e, a\}$;
(2) $e \circ x = x \circ e = x$, for every $x \in P$;
(3) $a \circ x = x \circ a = x$, for every $x \in P \setminus \{e, a\}$;
(4) $x \circ y = x \cdot y$, for every $x, y \in A_5$ such that $y \neq x^{-1}$;
(5) $x \circ x^{-1} = \{e, a\}$, for every $x \in P \setminus \{e, a\}$.

It is easy to see that $(P, \circ)$ is a polygroup. Moreover, $P/\beta^* \cong A_5$ and hence P is not solvable.

Chapter 4

Weak Polygroups

4.1 Weak hyperstructures

Weak hyperstructures or H_v-structures were introduced by Vougiouklis at the Fourth AHA congress (1990)[149]. The concept of an H_v-structure constitutes a generalization of the well-known algebraic hyperstructures (hypergroup, hyperring, hypermodule and so on). Actually some axioms concerning the above hyperstructures such as the associative law, the distributive law and so on are replaced by their corresponding weak axioms.

Since then the study of H_v-structure theory has been pursued in many directions by T. Vougiouklis, B. Davvaz, S. Spartalis, A. Dramalidis, Š. Hošková, and others.

In this section, firstly, we present some definitions and theorems on weak hyperstructures [143; 146].

Definition 4.1.1. Let H be a non-empty set and $\cdot : H \times H \longrightarrow \mathcal{P}^*(H)$ be a hyperoperation. The " $\cdot$ " in H is called *weak associative* if

$$x \cdot (y \cdot z) \cap (x \cdot y) \cdot z \neq \emptyset, \text{ for all } x, y, z \in H.$$

The " $\cdot$ " is called *weak commutative* if

$$x \cdot y \cap y \cdot x \neq \emptyset, \text{ for all } x, y \in H.$$

The " $\cdot$ " is called *strongly commutative* if

$$x \cdot y = y \cdot x, \text{ for all } x, y \in H.$$

The hyperstructure $(H, \cdot)$ is called an H_v-*semigroup* if " $\cdot$ " is weak associative. An H_v-semigroup is called an H_v-*group* if

$$a \cdot H = H \cdot a = H, \text{ for all } a \in H.$$

In an obvious way, the H_v-subgroup of an H_v-group is defined.

All the weak properties for hyperstructures can be applied for subsets.

For example, if $(H, \cdot)$ is a weak commutative H_v-group, then for all non-empty subsets A, B, C of H, we have

$$(A \cdot B) \cap (B \cdot A) \neq \emptyset \quad \text{and} \quad A \cdot (B \cdot C) \cap (A \cdot B) \cdot C \neq \emptyset.$$

To prove this, one has simply to take one element of each set.

Definition 4.1.2. Let $(H_1, \cdot)$, $(H_2, *)$ be two H_v-groups. A map $f : H_1 \longrightarrow H_2$ is called an H_v-*homomorphism* or a *weak homomorphism* if

$$f(x \cdot y) \cap f(x) * f(y) \neq \emptyset, \text{ for all } x, y \in H_1.$$

f is called an *inclusion homomorphism* if

$$f(x \cdot y) \subseteq f(x) * f(y), \text{ for all } x, y \in H_1.$$

Finally, f is called a *strong homomorphism* or a *good homomorphism* if

$$f(x \cdot y) = f(x) * f(y), \text{ for all } x, y \in H_1.$$

If f is onto, one to one and strong homomorphism, then it is called an *isomorphism*. Moreover, if the domain and the range of f are the same H_v-group, then the isomorphism is called an *automorphism*. We can easily verify that the set of all automorphisms of H, defined by $AutH$, is a group.

Several H_v-structures can be defined on a set H. A partial order on these hyperstructures can be introduced, as follows:

Definition 4.1.3. Let $(H, \cdot)$, $(H, *)$ be two H_v-groups defined on the same set H. We say that " $\cdot$ " less than or equal to " $*$ " and we write $\cdot \leq *$, if there is $f \in Aut(H, *)$ such that $x \cdot y \subseteq f(x * y)$, for all $x, y \in H$.

If a hyperoperation is weak associative, then every greater hyperoperation, defined on the same set is also weak associative. In [144], the set of all H_v-groups with a scalar unit defined on a set with three elements is determined using this property.

Theorem 4.1.4. *Greater hyperoperation from the one of a given H_v-group defines an H_v-group. The weak commutativity is also valid for every greater hyperoperation.*

Proof. Obvious from the definition. ■

We remark that the above theorem is not true for hypergroups.

Let $(H, \cdot)$ be an H_v-group. The relation β^* is the smallest equivalence relation on H such that the quotient H/β^* is a group. β^* is called the *fundamental equivalence relation* on H. If $\mathcal{U}$ denotes the set of all finite products of elements of H, then a relation β can be defined on H whose transitive closure is the fundamental relation β^*. The relation β is defined as follows: for x and y in H we write $x\beta y$ if and only if $\{x,y\} \subseteq u$ for some $u \in \mathcal{U}$. We can rewrite the definition of β^* on H as follows: $a\beta^*b$ if and only if there exist $z_1, \ldots, z_{n+1} \in H$ with $z_1 = a$, $z_{n+1} = b$ and $u_1, \ldots, u_n \in \mathcal{U}$ such that $\{z_i, z_{i+1}\} \subseteq u_i$ $(i = 1, \ldots, n)$. Suppose that $\beta^*(a)$ is the equivalence class containing $a \in H$. Then, the product $\odot$ on H/β^* is defined as follows:

$$\beta^*(a) \odot \beta^*(b) = \{\beta^*(c)|\ c \in \beta^*(a) \cdot \beta^*(b)\} \quad \text{for all}\ a, b \in H.$$

It is not difficult to see that $\beta^*(a) \odot \beta^*(b)$ is the singleton $\{\beta^*(c)\}$ for all $c \in \beta^*(a) \cdot \beta^*(b)$. In this way H/β^* becomes a group.

A motivation to study the above structures is given by the following examples:

Example 4.1.5.

(1) Let $(G, \cdot)$ be a group and R be an equivalence relation on G. In G/R consider the hyperoperation $\odot$ defined by $\overline{x} \odot \overline{y} = \{\overline{z}|\ z \in \overline{x} \cdot \overline{y}\}$, where $\overline{x}$ denotes the equivalence class of the element x. Then, $(G, \odot)$ is an H_v-group which is not always a hypergroup.

(2) On the set $\mathbb{Z}_{mn}$ consider the hyperoperation $\oplus$ defined by setting $0 \oplus m = \{0, m\}$ and $x \oplus y = x + y$ for all $(x, y) \in \mathbb{Z}_{mn}^2 - \{(0, m)\}$. Then, $(\mathbb{Z}_{mn}, \oplus)$ is an H_v-group. $\oplus$ is weak associative but not associative, since taking $k \notin m\mathbb{Z}$ we have

$$(0 \oplus m) \oplus k = \{0, m\} \oplus k = \{k, m + k\},$$
$$0 \oplus (m \oplus k) = 0 \oplus (m + k) = \{m + k\}.$$

Moreover, it is weak commutative but not commutative. From the sum

$$[\ldots (0 \oplus m) \oplus \underbrace{\ldots \oplus m] \oplus m}_{n-1 \text{ times}} = \{0, m, 2m, \ldots, (n - 1)m\},$$

it is obtained that $\beta(0) = \{0, m, 2m, \ldots, (n-1)m\}$. Similarly, for every $0 < k < n - 1$ we have

$$k \oplus [\ldots (0 \oplus m) \oplus \underbrace{\ldots \oplus m] \oplus m}_{n-1 \text{ times}} = \{k, k+m, k+2m, \ldots, k+(n-1)m\}.$$

So $\beta(k) = k + m\mathbb{Z}$. That means that $\beta^* = \beta$ and $\mathbb{Z}_{mn}/\beta^* \cong \mathbb{Z}_m$.

(3) Consider the group $(\mathbb{Z}^n, +)$ and take $m_1, \ldots, m_n \in \mathbb{N}$. We define a hyperoperation $\oplus$ in $\mathbb{Z}^n$ as follows:

$$(m_1, 0, \ldots, 0) \oplus (0, 0, \ldots, 0) = \{(m_1, 0, \ldots, 0), (0, 0, \ldots, 0)\},$$
$$(0, m_2, \ldots, 0) \oplus (0, 0, \ldots, 0) = \{(0, m_2, \ldots, 0), (0, 0, \ldots, 0)\},$$
$$(0, 0, \ldots, m_n) \oplus (0, 0, \ldots, 0) = \{(0, 0, \ldots, m_n), (0, 0, \ldots, 0)\},$$

and $\oplus = +$ in the remaining cases. Then, $(\mathbb{Z}^n, \oplus)$ is an H_v-group and we have

$$\mathbb{Z}^n / \beta^* \cong \mathbb{Z}_{m_1} \times \mathbb{Z}_{m_2} \times \ldots \times \mathbb{Z}_{m_n}.$$

Let $(H, \cdot)$ be an H_v-group. An element $x \in H$ is called *single* if its fundamental class is singleton, i.e., $\beta^*(x) = \{x\}$. Denote by S_H the set of all single elements of H.

Theorem 4.1.6. *Let $(H, \cdot)$ be an H_v-group and $x \in S_H$. Let $a \in H$ and take any element $v \in H$ such that $x \in a \cdot v$. Then,*

$$\beta^*(a) = \{h \in H \mid h \cdot v = x\}.$$

Proof. We have $x \in a \cdot v$. So, $x = a \cdot v$ which means $x = \beta^*(a) \cdot \beta^*(v)$. Thus, for all $h \in \beta^*(a)$ we have $h \cdot v = x$. Conversely, let $x = h \cdot v$. Then, $x = \beta^*(h) \cdot \beta^*(v)$. Since H/β^* is a group, we have $\beta^*(h) = x \cdot (\beta^*(v))^{-1} = \beta^*(a)$, so $h \in \beta^*(a)$. ∎

Theorem 4.1.7. *Let $(H, \cdot)$ be an H_v-group and $x \in S_H$. Then, the core of H is $\omega_H = \{u \mid u \cdot x = x\} = \{u \mid x \cdot u = x\}$.*

Proof. It is obvious. ∎

Theorem 4.1.8. *Let $(H, \cdot)$ be an H_v-group and $x \in S_H$. Then, $x \cdot y = \beta^*(x \cdot y)$ and $y \cdot x = \beta^*(y \cdot x)$, for all $y \in H$.*

Proof. Suppose that for some y there exist $t \in x \cdot y$ and $t' \in \beta^*(t)$ such that $t' \notin x \cdot y$. From the reproductivity, there exists $v \neq x$ in H such that $t' \in v \cdot y$. So, we have $\beta^*(v) \cdot \beta^*(y) = \beta^*(t')$. On the other hand, $\beta^*(x) \cdot \beta^*(y) = \beta^*(t)$. Thus, $\beta^*(v) \cdot \beta^*(y) = \beta^*(x) \cdot \beta^*(y)$. Hence, $\beta^*(v) = \beta^*(x)$. Since $x \in S_H$, $\beta^*(x) = \{x\}$. Thus, $v = x$ which is a contradiction. ∎

The previous theorem proves that the product of a single element with any arbitrary element is always a whole fundamental class.

Suppose that $(H, \cdot)$ is an H_v-group such that S_H is non-empty. Then, the only greater hyperoperations $\cdot < *$ for which the H_v-groups $(H, *)$ contain single elements are the ones with the same fundamental group, since the fundamental classes are determined from the products of a single element with the elements of the group. On the other hand, a less hyperoperation $\circ < \cdot$ can have the same set S_H if only in the products of non-single elements the $\circ$ is less than $\cdot$. Finally, if ρ and σ are equivalence relations with $\rho < \sigma$ such that H/ρ and H/σ are non-equal groups, then they can not have both single elements.

Corollary 4.1.9. *Let* $(H, \cdot)$ *be an* H_v-group. *If* S_H *is non-empty, then* $\beta^* = \beta$.

Proof. It is obvious, since all the β^*-classes can be obtained as products of two elements one of which is single. ■

Let $(H, \cdot)$ be an H_v-group with (left, right) identity elements. Then, H is called (*left, right*) *reversible* in itself when any relation $c \in a \cdot b$ implies the existence of a left inverse a' of a and a right inverse b' of b such that $b \in a' \cdot c$ and $a \in c \cdot b'$.

An H_v-group $(H, \cdot)$ is called *feebly quasi-canonical* if it is regular, reversible and satisfies the following conditions:

> For each $a \in H$, if a', a'' are inverses of a, then for each $x \in H$, we have $a' \cdot x = a'' \cdot x$ and $x \cdot a' = x \cdot a''$.

A feebly quasi-canonical H_v-group H is called *feebly canonical* if it is strongly commutative.

In the rest of this section, we study a wide class of reversible H_v-groups that investigated by Spartalis [136].

Let $(H, \cdot)$ be an H_v-group with left or right identity elements. We denote by E_l (respectively, E_r) the set of left (respectively, right) identities. Hence, the set of all identities of the H_v-group H is $E = E_l \cup E_r$. We also denote by $i_l(x, e)$ the set of all left inverses of an element x with respect to the identity e of E, i.e., $i_l(x, e) = \{x' \in H \mid e \in x' \cdot x\}$. Consequently, $i_l(x) = \bigcup_{e \in E} i_l(x, e)$ is the set of all left inverses of the element x. If A is a non-empty subset of H, then $i_l(A) = \bigcup_{a \in A} i_l(a)$. Similar notations hold for the right inverses, too. The following definition is similar to Definition 2.6.2.

Definition 4.1.10. Let $(H, \cdot)$ be an H_v-group. Then, H is called *left completely reversible* in itself if, for all $a, b \in H$, it satisfies the following condition:

(1) $c \in a \cdot b$ implies $b \in u \cdot c$, for all $u \in i_l(a)$.

Similarly, H is called *right completely reversible* in itself if

(2) $c \in a \cdot b$ implies $a \in c \cdot v$, for all $v \in i_r(b)$.

Lemma 4.1.11. *If H is left completely reversible, then for each $e_l \in E_l$, $e_r \in E_r$, $e \in E = E_l \cup E_r$ and $a \in H$, we have*

(1) $i_l(e_l) = E_l$;
(2) $i_l(a) = i_l(a, e_r)$;
(3) $i_l(i_l(i_l(a, e))) \subseteq i_l(a, e)$.

Proof. (1) Suppose that $e_l \in E_l$. Clearly, $E_l \subseteq i_l(e_l, e_l) \subseteq i_l(e_l)$. Moreover, let $u \in i_l(e_l)$. Since for all $x \in H$, $x \in e_l \cdot x$, it follows that $x \in u \cdot x$. Therefore, $u \in E_l$, i.e., $i_l(e_l) \subseteq E_l$ and so $i_l(e_l) = E_l$.

(2) Suppose that $a \in H$ and $e_r \in E_r$. Obviously, $i_l(a, e_r) \subseteq i_l(a)$. Conversely, assume that $u \in i_l(a)$. From the relation $a \in a \cdot e_r$ it follows that $e_r \in u \cdot a$, that is $u \in i_l(a, e_r)$ and so $i_l(a) = i_l(a, e_r)$.

(3) Suppose that $a \in H$, $e \in E$ and $u \in i_l(a, e)$. Since $e \in u \cdot a$, we have that $a \in w \cdot e$ for all $w \in i_l(u)$ and $e \in v \cdot a$ for all $v \in i_l(i_l(u))$. Consequently, $v \in i_l(a, e)$ and so $i_l(i_l(i_l(a, e))) \subseteq i_l(a, e)$. ∎

Proposition 4.1.12. *Let H be left completely reversible, $a \in H$ and $e \in E$ such that $a \in i_l(i_l(a, e))$. Then, the following conditions hold:*

(1) $i_l(a, e) = i_l(a) = i_l(i_l(i_l(a, e)))$;
(2) If u is an inverse of a with respect to e for which the hypothesis holds, i.e., $a \in i_l(u)$, then $i_l(i_l(a)) = i_l(u)$ and, moreover, for all $x \in H$, $v \in i_l(a)$ we have $u \cdot x \subseteq v \cdot x$.

Proof. (1) Suppose that $a \in i_l(i_l(a, e))$. Then, $i_l(a) \subseteq i_l(i_l(i_l(a, e)))$. According to Lemma 4.1.11 (3), we have

$$i_l(a) \subseteq i_l(i_l(i_l(a, e))) \subseteq i_l(a, e) \subseteq i_l(a).$$

Thus, $i_l(a, e) = i_l(a) = i_l(i_l(i_l(a, e)))$.

(2) Suppose that u is an inverse of the element a with respect to e such that $a \in i_l(u)$. Then, there exists an identity $e' \in E$ such that $a \in i_l(u, e')$. Therefore, $u \in i_l(a) \subseteq i_l(i_l(u, e'))$. According to (1), we have $i_l(u, e') = i_l(u) = i_l(i_l(i_l(u)))$. Consequently, $u \in i_l(a) \subseteq i_l(i_l(u))$. So, $i_l(u) \subseteq i_l(i_l(a)) \subseteq i_l(u)$. Hence, $i_l(i_l(a)) = i_l(u)$. Finally, let $x \in H$ and $y \in u \cdot x$. From the reversibility of H, it follows that $x \in a \cdot y$ and $y \in v \cdot x$, for all $v \in i_l(a)$. Thus, $u \cdot x \subseteq v \cdot x$. ∎

Notice that if $a \in H$, $a \in E$ and $a \in i_l(a, e)$, then the following relation is satisfied:

$$i_l(a, e) = i_l(a) = i_l(i_l(i_l(a, e))) = i_l(i_l(a)).$$

Proposition 4.1.13. *Let H be left completely reversible and $a \in H$ such that $a \in \bigcap_{u \in i_l(a)} i_l(u)$. Then, for all $x \in H$, $u, v \in i_l(a)$ and $e_r \in E_r$ we have*

(1) $u \cdot x = v \cdot x$;
(2) If E_r is non-empty, then $i_l(a) \subseteq u \cdot e_r \subseteq i_r(a, e_r)$.

Proof. (1) We observe that the assumption of Proposition 4.1.12 are satisfied for all $e \in E$ and for all $u \in i_l(a)$. Therefore, for all $x \in H$ and $u, v \in i_l(a)$, we have $u \cdot x \subseteq v \cdot x$ and so $u \cdot x = v \cdot x$.

(2) Suppose that $e_r \in E_r$. Then, we have $i_l(a) \cdot e_r = \bigcup_{u \in i_l(a)} u \cdot e_r$ and because of (1), it follows that $i_l(a) \cdot e_r = u \cdot e_r$, for all $u \in i_l(a)$. Thus, $i_l(a) \subseteq u \cdot e_r$. Moreover, let $y \in u \cdot e_r$. Since $a \in i_l(u)$, it follows that $e_r \in a \cdot y$, that is, $y \in i_r(a, e_r)$. ■

If $(H, \cdot)$ is a strongly commutative H_v-group, then from the first condition of previous proposition we conclude that the concepts of left (respectively, right) completely reversible H_v-group and the feebly canonical H_v-group are identical.

Theorem 4.1.14. *If H is completely reversible, then it is a feebly quasi-canonical H_v-group.*

Proof. At the first, we prove that all identities and inverses in H are two sided. Let $a \in H$, $e \in E$, $u \in i_l(a, e)$ and $v \in i_r(a, e)$. Since $e \in a \cdot v$, it follows that $v \in u \cdot e$. Since $e \in i_r(e)$, $u \in v \cdot e$. Since $a \in i_l(v, e)$, we have $e \in a \cdot u$. Therefore, $i_l(a, e) \subseteq i_r(a, e)$. Similarly, we obtain $i_r(a, e) \subseteq i_l(a, e)$. Hence, $i_l(a, e) = i_r(a, e)$. Now, suppose that $e_l \in E_l$ and $a \in H$. Then, there exists $u \in H$ such that $e_l \in u \cdot a \cap a \cdot u$ and so $a \in a \cdot e_l$. Thus, $E_l \subseteq E_r$. In the same manner, every right identity is also a left identity and hence $E_l = E_r$. Consequently, H is regular and reversible. Moreover, by Proposition 4.1.13 (1) for the left (right) completely reversible H_v-groups, we have that, for $a \in H$, if u, v are inverses of a, then for all $x \in H$, $u \cdot x = v \cdot x$ and $x \cdot u = x \cdot v$. Therefore, H is a feebly quasi-canonical H_v-group. ■

In what follows, we consider H_v-groups with only two sided identities and inverses. The following relation introduced by De Salvo [66] and studied by Spartalis [136].

Consider the binary relation $\sim$ on H as follows:

$$x \sim y \ \Leftrightarrow \ \text{there exists } z \in H \text{ such that } \{x, y\} \subseteq i(z),$$

for all $x, y \in H$. If H is left completely reversible, then by using Proposition 4.1.13 (1) we obtain that $\sim$ is an equivalence relation. In the quotient set $\hat{H} = H/\sim$ we define the following hyperoperation between classes in the usual manner

$$\hat{x} \odot \hat{y} = \{\hat{w} \mid w \in \hat{x} \cdot \hat{y}\},$$

for all $x, y \in H$, where $\hat{x}$ is the equivalence class containing x. According to Proposition 4.1.13 (1), this hyperoperation is equivalent to the following one

$$\hat{x} \odot \hat{y} = \{\hat{w} \mid w \in x \cdot y\}.$$

Lemma 4.1.15. $(\hat{H}, \odot)$ *is an H_v-group.*

Proof. It is straightforward. ∎

Proposition 4.1.16. *Let H be left completely reversible and $e \in E$. Then, $\hat{H}$ is a left reversible in itself H_v-group and $\hat{e} = E$ is the unique identity of $\hat{H}$. Moreover, $\hat{e}$ is a right scalar, $\hat{H}$ is regular and each $\hat{x} \in \hat{H}$ has a unique inverse.*

Proof. Suppose that $a \in E$. By Lemma 4.1.11 (1) we have $i(e) = E$. So, $E \subseteq \hat{e}$. Further, if $x \in \hat{e}$, then there exists $y \in H$ such that $\{x, e\} \subseteq i(y)$. Since $y \in i(e)$, it follows that $x \in E$. Hence, $\hat{e} \subseteq E$. Thus, $\hat{e} = E$. Obviously, for all $\hat{x} \in \hat{H}$, $\hat{x} \in \hat{x} \odot \hat{e} \cap \hat{e} \odot \hat{x}$. Now, suppose that $\hat{s}$ is a left identity or $\hat{t}$ is a right identity of $\hat{H}$. Then, $\hat{e} \in \hat{s} \odot \hat{e} \cap \hat{e} \odot \hat{t}$ and so there exist $e' \in E$ and $t' \in \hat{t}$ such that $e \in s \cdot e' \cap e \cdot t'$. Therefore, $s, t' \in i(e) = i(e')$ and hence $\hat{s} = \hat{e} = \hat{t}$. Finally, for all $\hat{x} \in \hat{H}$,

$$\hat{x} \odot \hat{e} = \{\hat{w} \mid w \in x \cdot e = x \cdot E\}.$$

If $w \in x \cdot e'$ and $e' \in E$, then for all $u \in i(x)$, $e' \in u \cdot w$. Thus, $\{w, x\} \subseteq i(u)$ and so $x \sim w$. Consequently, $\hat{x} \odot \hat{e} = \{\hat{x}\}$.

It is easy to prove that each $\hat{x} \in \hat{H}$ has a unique inverse, i.e., $i(\hat{x}) = \{\hat{a}\}$, where $a \in i(x)$. Moreover, $\hat{H}$ is a regular H_v-group. Finally, we show that $\hat{H}$ is left reversible in itself. Suppose that $\hat{x}, \hat{y} \in \hat{H}$ and $\hat{z} \in \hat{x} \odot \hat{y}$. Then, there exists $y' \in \hat{y}$ such that $z \in x \cdot y'$ and hence for all $a \in i(x)$, $y' \in a \cdot z$. Since $i(\hat{x}) = \{\hat{a}\}$, it follows that $\hat{y} \in \hat{a} \odot \hat{z}$. ∎

Now, let K be an H_v-subgroup of H. Let the left coset expansion $H/K = \{x \cdot K \mid x \in H\}$ satisfies the following conditions:

(1*) for all $x \in H$, $x \in x \cdot K$,

(2*) for all $x, y \in H$, $x \cdot K \cap y \cdot K \neq \emptyset$ implies $x \cdot K = y \cdot K$.

It is easy to see that H/K becomes an H_v-group with respect to the usual hyperoperation:

$$x \cdot K \odot y \cdot K = \{z \cdot K \mid z \in (x \cdot K) \cdot (y \cdot K)\},$$

for all $x, y \in H$. A similar remark holds for the right coset expansion. Moreover, for the right coset expansion, we have the following proposition.

Proposition 4.1.17. *Let H be left completely reversible and K be an H_v-subgroup of K. If $K \cap E \neq \emptyset$ and for all $x \in H$, $K \cdot (K \cdot x) \subseteq K \cdot x$, then $H/K = \{K \cdot x \mid x \in H\}$ is an H_v-group.*

Proof. It suffices to prove the following conditions:

(1) for all $x \in H$, $x \in K \cdot x$,

(2) for all $x, y \in H$, $K \cdot x \cap K \cdot y \neq \emptyset$ implies $K \cdot x = K \cdot y$.

Obviously, for all $x \in H$, $x \in K \cdot x$. Moreover, suppose that $x, y \in H$ and $z \in K \cdot x \cap K \cdot y$. Then, $K \cdot z \subseteq K \cdot (K \cdot x) \subseteq K \cdot x$. From $z \in K \cdot x$, it follows that $z \in u \cdot x$, where $u \in K$. Therefore, for all $a \in i(u)$, $x \in a \cdot z$. Since $i(u) \cap K \neq \emptyset$, we have $x \in K \cdot z$. Thus, $K \cdot x \subseteq K \cdot z$. Consequently, $K \cdot z = K \cdot x$. Similarly, $K \cdot z = K \cdot y$. ∎

Theorem 4.1.18. *Let H be left completely reversible and E be the set of identities and H/E be the left coset expansion of H with respect to E. Then the following condition hold:*

(1) *E is a total H_v-subgroup of H;*

(2) *for all $x \in H$, $x \cdot E = \hat{x}$, that is H/E is identical with the H_v-group $\hat{H}$;*

(3) *E is the smallest of the H_v-subgroup K of H such that H/K satisfies (1*) and (2*).*

Proof. By using Proposition 4.1.13 (2) and Lemma 4.1.11 (2), for all $a \in H$, $e \in E$ and $u \in i(a)$, we obtain

$$i(a) = u \cdot a = i(a, e). \qquad (\star)$$

(1) Suppose that $e \in E$. Then, from the previous relation, we obtain

$$i(e) = u \cdot e', \text{ for all } u \in i(e) \text{ and } e' \in E.$$

Moreover, according to Lemma 4.1.11 (1), $i(e) = E$ and hence $E = e'' \cdot e'$, for all $e', e'' \in E$. Therefore, E is a total H_v-subgroup of H.

(2) By hypothesis $H/E = \{x \cdot E \mid x \in H\}$ and $x \cdot E = \bigcup_{e \in E} x \cdot e$. Suppose that $x \in H$, $e \in E$ and $u \in i(x)$. Then, applying $(\star)$ we obtain $x \cdot E = i(u) = x \cdot e$ and hence $H/E = \{x \cdot e \mid x \in H\}$. Furtheremore, for all $x, y \in H$ we have the following

$$x \sim y \Rightarrow \text{there exists } z \in H \text{ such that } \{x, y\} \subseteq i(z)$$
$$\Leftrightarrow \text{there exists } z \in H, \ e \in E \text{ such that } y \in x \cdot e = i(z)$$
$$\Leftrightarrow y \in x \cdot e.$$

Consequently, H/E is identical with the H_v-group $\hat{H}$.

(3) Suppose that K is an H_v-subgroup of H such that the left coset expansion H/K satisfies (1). Let $e \in E$. Then, $e \in e \cdot K$ and so $e \in e \cdot u$, for some $u \in K$. Therefore, $u \in i(e)$ and since $i(e) = E$ we have $K \cap E \neq \emptyset$. Finally, since E is a total H_v-subgroup of H, it follows that $E \subseteq K$. ∎

4.2   Weak polygroups as a generalization of polygroups

In this section, we study the concept of weak polygroups which is a generalization of polygroups [46; 50].

Definition 4.2.1.   A multivalued system $\mathcal{M} =< P, \cdot, e, ^{-1} >$, where $e \in P$, $^{-1} : P \longrightarrow P$, $\cdot : P \times P \longrightarrow \mathcal{P}^*(P)$ is called a *weak polygroup* if the following axioms hold for all $x, y, z \in P$:

(1) $(x \cdot y) \cdot z \cap x \cdot (y \cdot z) \neq \emptyset$ (*weak associative*),
(2) $x \cdot e = x = e \cdot x$,
(3) $x \in y \cdot z$ implies $y \in x \cdot z^{-1}$ and $z \in y^{-1} \cdot x$.

The following elementary facts about weak polygroups follow easily from the axioms: $e \in x \cdot x^{-1} \cap x^{-1} \cdot x$, $e^{-1} = e$, $(x^{-1})^{-1} = x$.

Example 4.2.2. Consider $P = \{e, a, b, c\}$ and define $*$ on P with the help of the following table:

$*$	e	a	b	c
e	e	a	b	c
a	a	$\{e,a\}$	c	b
b	b	c	$\{e,b\}$	a
c	c	b	a	$\{e,c\}$

Then, $< P, *, e, ^{-1} >$, where $x^{-1} = x$ for every $x \in P$, is a weak polygroup which is not a polygroup. Indeed, we have

$$(a * b) * c = c * c = \{e, c\} \text{ and } a * (b * c) = a * a = \{e, a\}.$$

Therefore, $*$ is not associative.

Proposition 4.2.3. *Let* $(G, \cdot)$ *be a group and* θ *be an equivalence relation on* G *such that*

(1) $x\theta e$ *implies* $x = e$,
(2) $x\theta y$ *implies* $x^{-1}\theta y^{-1}$.

Let $\theta(x)$ *be the equivalence class of the element* $x \in G$. *If* $G/\theta = \{\theta(x) \mid x \in G\}$, *then* $< G/\theta, \odot, \theta(e), ^{-I} >$ *is a weak polygroup, where the hyperoperation* $\odot$ *is defined as follows:*

$$\odot : G/\theta \times G/\theta \to \mathcal{P}^*(G/\theta)$$
$$\theta(x) \odot \theta(y) = \{\theta(z) | z \in \theta(x).\theta(y)\},$$

and $\theta(x)^{-I} = \theta(x^{-1})$.

Proof. For all $x, y, z \in G$, we have

$$x \cdot (y \cdot z) \in \theta(x) \odot (\theta(y) \odot \theta(z)),$$
$$(x \cdot y) \cdot z \in (\theta(x) \odot \theta(y)) \odot \theta(z).$$

Therefore, $\odot$ is weak associative. It is easy to see that $\theta(e)$ is the identity element in G/θ and $\theta(x^{-1})$ is the inverse of $\theta(x)$ in G/θ. Now, we show that:

$\theta(z) \in \theta(x) \odot \theta(y)$ implies $\theta(x) \in \theta(z) \odot \theta(y^{-1})$ and $\theta(y) \in \theta(x^{-1}) \odot \theta(z)$.

We have $\theta(z) \in \theta(x) \odot \theta(y) = \{\theta(a) \mid a \in \theta(x).\theta(y)\}$. Hence, $\theta(z) = \theta(a)$ for some $a \in \theta(x).\theta(y)$. Therefore, there exist $b \in \theta(x)$ and $c \in \theta(y)$ such that $a = b \cdot c$, so $b = a \cdot c^{-1}$ which implies that $\theta(b) = \theta(a.c^{-1}) \in \theta(a) \odot \theta(c^{-1})$. Therefore,

$$\theta(x) \in \theta(z) \odot \theta(y^{-1}).$$

By a similar way, we obtain

$$\theta(y) \in \theta(x^{-1}) \odot \theta(z).$$

Therefore, $< G/\theta, \odot, \theta(e), ^{-I} >$ is a weak polygroup. ■

An extension of polygroups by polygroups introduced in Section 3.2. We can consider an extension of a weak polygroup by another weak polygroup in a similar way.

Theorem 4.2.4. *Let* $\mathcal{A} =< A, \cdot, e, ^{-1} >$ *and* $\mathcal{B} =< B, \cdot, e, ^{-1} >$ *be weak polygroups whose elements have been renamed so that* $A \cap B = \{e\}$, *where* e *is the identity of both* $\mathcal{A}$ *and* $\mathcal{B}$. *A new system* $\mathcal{A}[\mathcal{B}] =< M, *, e, ^{-1} >$, *which is called the extension of* $\mathcal{A}$ *by* $\mathcal{B}$ *is a weak polygroup.*

Proof. The verification of the third condition of Definition 4.2.1 is similar to the proof of Theorem 3.2.4. Therefore, we show that weak associativity is valid. For all x, y, z in M, we consider the following cases:

(1) If $x, y, z \in A$, then $(x \cdot y) \cdot z = (x * y) * z$ and $x \cdot (y \cdot z) = x * (y * z)$,

(2) If $x, y, z \in B$, then $(x \cdot y) \cdot z \subseteq (x * y) * z$ and $x \cdot (y \cdot z) \subseteq x * (y * z)$,

(3) If $x \in A, y, z \in B$, then $(y \cdot z) \subseteq (x * y) * z$ and $y \cdot z \subseteq x * (y * z)$,

(4) If $x \in A, y \in B, z \in A$, then $y \in (x * y) * z$ and $y \in x * (y * z)$,

(5) If $x \in A, y \in A, z \in B$, then $z \in (x * y) * z$ and $z \in x * (y * z)$,

(6) If $x \in B, y, z \in A$, then $x \in (x * y) * z$ and $x \in x * (y * z)$,

(7) If $x \in B, y \in B, z \in A$, then $x \cdot y \subseteq (x * y) * z$ and $x \cdot y \subseteq x * (y * z)$,

(8) If $x \in B, y \in A, z \in B$, then $x \cdot z \subseteq (x * y) * z$ and $x \cdot z \subseteq x * (y * z)$.

Thus, $*$ is weak associative. ∎

The following definition, first defined in Section 3.3 for polygroups.

The equivalence relation θ on a weak polygroup $\mathcal{M}$ is called a *(full) conjugation* on $\mathcal{M}$ if

(1) $x\theta y$ implies $x^{-1}\theta y^{-1}$,

(2) $z \in x \cdot y$ and $z_1 \theta z$ imply $z_1 \in x_1 \cdot y_1$ for some x_1 and y_1 where $\theta(x_1) = \theta(x)$ and $\theta(y_1) = \theta(y)$.

The collection of all θ-classes, with the induced operation from $\mathcal{M}$, forms a weak polygroup.

Corollary 4.2.5. *Let* $\mathcal{M}$ *be a weak polygroup, then* θ *is a conjugation on* $\mathcal{M}$ *if and only if*

(1) $(\theta(x))^{-1} = \theta(x^{-1})$;

(2) $\theta(\theta(x)y) = \theta(x)\theta(y)$.

Proof. The proof is similar to the proof of Lemma 3.3.11. ∎

Definition 4.2.6. Let $\mathcal{A} =< A, ., e_1, ^{-1} >$ and $\mathcal{B} =< B, *, e_2, ^{-1} >$ be two weak polygroups, and let f be a mapping from A into B, such that $f(e_1) = e_2$. Then, f is called

(1) a *weak homomorphism*, if $f(x \cdot y) \cap f(x) * f(y) \neq \emptyset$, for all $x, y \in A$,

(2) an *inclusion homomorphism*, if $f(x \cdot y) \subseteq f(x) * f(y)$, for all $x, y \in A$,

(3) a *strong homomorphism*, if $f(x \cdot y) = f(x) * f(y)$, for all $x, y \in A$.

If f is one to one, onto and strong homomorphism, then it is called an *isomorphism*. Moreover, if f is defined on the same weak polygroup, then it is called an *automorphism*. The set of all automorphism of A, written $Aut A$, is a group.

Definition 4.2.7. If A is a subpolygroup of a weak polygroup P, then we define the relation $a \equiv b \ (mod A)$ if and only if there exists a set $\{c_0, c_1, \ldots, c_{k+1}\} \subseteq P$, where $c_0 = a$, $c_{k+1} = b$ such that

$$a \cdot c_1^{-1} \cap A \neq \emptyset, \quad c_1 \cdot c_2^{-1} \cap A \neq \emptyset, \quad \ldots, \quad c_k \cdot b^{-1} \cap A \neq \emptyset.$$

This relation is denoted by $a A_P^* b$.

Lemma 4.2.8. *The relation A_P^* is an equivalence relation.*

Proof. (1) Since $e \in a \cdot a^{-1} \cap A$ for all $a \in P$; then $a A_P^* a$, i.e., A_P^* is reflexive.

(2) Suppose that $a A_P^* b$. Then, there exists $\{c_0, c_1, \ldots, c_{k+1}\} \subseteq P$ where $c_0 = a$, $c_{k+1} = b$ such that

$$a \cdot c_1^{-1} \cap A \neq \emptyset, \quad c_1 \cdot c_2^{-1} \cap A \neq \emptyset, \quad \ldots, \quad c_k \cdot b^{-1} \cap A \neq \emptyset.$$

Therefore, there exists $x_i \in c_i \cdot c_{i+1}^{-1} \cap A \quad (i = 0, \ldots, k)$ which implies that $x_i^{-1} \in c_{i+1} \cdot c_i^{-1}$ and $x_i^{-1} \in A$, this means that $b A_P^* a$, and so A_P^* is symmetric.

(3) Let $a A_P^* b$ and $b A_P^* c$, where $a, b, c \in P$. Then, there exist

$$\{c_0, c_1, \ldots, c_{k+1}\} \subseteq P \text{ and } \{d_0, d_1, \ldots, d_{r+1}\} \subseteq P,$$

where $c_0 = a$, $c_{k+1} = b = d_0$, $d_{r+1} = c$ such that

$$a \cdot c_1^{-1} \cap A \neq \emptyset, \ c_1 \cdot c_2^{-1} \cap A \neq \emptyset, \ldots, c_k \cdot b^{-1} \cap A \neq \emptyset,$$

$$b \cdot d_1^{-1} \cap A \neq \emptyset, \ d_1 \cdot d_2^{-1} \cap A \neq \emptyset, \ldots, d_r \cdot c^{-1} \cap A \neq \emptyset.$$

We take $\{c_0, c_1, \ldots, c_{k+1}, d_1, d_2, \ldots, d_{r+1}\} \subseteq P$ which satisfies the condition for $a A_P^* c$, and so A_P^* is transitive. Therefore, A_P^* is an equivalence relation. ■

We denote $A_P^*[x]$ the equivalence class with representative x.

Theorem 4.2.9. *Let P be a weak polygroup. If A is a subpolygroup of P, then on the set $[P : A] = \{A_P^*[a] \mid a \in P\}$ we define the hyperoperation $\odot$ as follows:*

$$A_P^*[a] \odot A_P^*[b] = \{A_P^*[c] \mid c \in A_P^*[a] \cdot A_P^*[b]\},$$

what gives the weak polygroup $< [P : A], \odot, A_P^*[e], ^{-I} >$, *where* $A_P^*[a]^{-I} = A_P^*[a^{-1}]$.

Proof. For $a, b, c \in P$, we have

$$(a \cdot b) \cdot c \subseteq (A_P^*[a] \cdot A_P^*[b]) \cdot A_P^*[c],$$
$$a \cdot (b \cdot c) \subseteq A_P^*[a] \cdot (A_P^*[b] \cdot A_P^*[c]).$$

Thus, $\odot$ is weak associative. Now, we show that A is the unit element in $[P : A]$. Obviously, we have $A \subseteq A_P^*[e]$. On the other hand, if $a \in A_P^*[e]$, then there exists a set $\{c_0, c_1, \ldots, c_{n+1}\} \subseteq P$ where $c_0 = a, c_{n+1} = e$ such that

$$a \cdot c_1^{-1} \cap A \neq \emptyset, \quad c_1 \cdot c_2^{-1} \cap A \neq \emptyset, \quad \ldots, \quad c_n \cdot e^{-1} \cap A \neq \emptyset.$$

So $c_n \in A$. Since $c_{n-1} \cdot c_n^{-1} \cap A \neq \emptyset$, there exists $x \in c_{n-1} \cdot c_n^{-1} \cap A$ which implies $c_{n-1} \in x \cdot c_n$, and so $c_{n-1} \in A$. By induction, we obtain $a \in A$. Therefore, $A_P^*[e] = A$. Now, we show that $A_P^*[a] \odot A_P^*[e] = A_P^*[a]$. Suppose that $A_P^*[z] \in A_P^*[a] \odot A_P^*[e]$. We claim that $A_P^*[z] = A_P^*[a]$. We have $z \in A_P^*[a] \cdot A_P^*[e]$. Hence, there exist $x \in A_P^*[a]$ and $y \in A$ such that $z \in x \cdot y$ which implies that $y \in x^{-1} \cdot z$. Then, $x^{-1} \cdot z \cap A \neq \emptyset$, and so $A_P^*[x] = A_P^*[z]$. Therefore, $A_P^*[z] = A_P^*[a]$. It is easy to see that $A_P^*[a^{-1}]$ is the inverse of $A_P^*[a]$ in $[P : A]$. Now, we show that $A_P^*[c] \in A_P^*[a] \odot A_P^*[b]$ implies $A_P^*[a] \in A_P^*[c] \odot A_P^*[b^{-1}]$ and $A_P^*[b] \in A_P^*[a^{-1}] \odot A_P^*[c]$. Since $A_P^*[c] \in A_P^*[a] \odot A_P^*[b]$, we have $A_P^*[c] = A_P^*[x]$ for some $x \in A_P^*[a] \cdot A_P^*[b]$. Therefore, there exist $y \in A_P^*[a]$ and $z \in A_P^*[b]$ such that $x \in y \cdot z$, so $y \in x \cdot z^{-1}$. This implies that $A_P^*[y] \in A_P^*[x] \odot A_P^*[z^{-1}]$, and so $A_P^*[a] \in A_P^*[c] \odot A_P^*[b^{-1}]$. Similarly, we get $A_P^*[b] \in A_P^*[a^{-1}] \odot A_P^*[c]$. Therefore, $< [P : A], \odot, A, ^{-I} >$ is a weak polygroup. ∎

If A is a subpolygroup of a weak polygroup P, then the weak polygroup $[P : A]$, is called the *quotient weak polygroup* of P by A.

Corollary 4.2.10. *Let ρ be a strong homomorphism from a weak polygroup P_1 into a weak polygroup P_2. Then, the following propositions hold:*

(1) *For all $a \in P_1$, $\rho(a^{-1}) = \rho(a)^{-1}$;*

(2) *The kernel of ρ is a subpolygroup of P_1;*

(3) *Let A be a subpolygroup of P_1. The image $\rho(A) = \{\rho(x) \mid x \in A\}$ is a subpolygroup of P_2. For a subpolygroup B of P_2, the inverse image $\rho^{-1}(B) = \{x \mid x \in P_1, \rho(x) \in B\}$ is a subpolygroup of P_1.*

Let P_1, P_2 be two weak polygroups and ρ a strong homomorphism of P_1 onto P_2. If K is the kernel of ρ then we can form $[P_1 : K]$. It is fairly natural to expect that there should be a very close relationship between P_2 and $[P_1 : K]$. The fundamental homomorphism theorem, which we are about to prove, spells out this relationship in exact detail.

Theorem 4.2.11. (Fundamental Homomorphism Theorem). *Let ρ be a strong homomorphism from P_1 onto P_2 with kernel K. Then,*

$$[P_1 : K] \cong P_2.$$

Proof. We define $\varphi : [P_1 : K] \longrightarrow P_2$ as follows:

$$\varphi(K^*_{P_1}[x]) = \rho(x), \text{ for all } x \in P_1.$$

This mapping is well defined, because if $K^*_{P_1}[x] = K^*_{P_1}[y]$, then there exists $\{z_0, z_1, \ldots, z_{k+1}\} \subseteq P_1$ where $z_0 = x$, $z_{k+1} = y$ such that

$$x \cdot z_1^{-1} \cap K \neq \emptyset, \ z_1 \cdot z_2^{-1} \cap K \neq \emptyset, \ \ldots, \ z_k \cdot y^{-1} \cap K \neq \emptyset.$$

Thus, $e_2 \in \rho(x \cdot z_1^{-1})$, $e_2 \in \rho(z_1 \cdot z_2^{-1})$, $\ldots$, $e_2 \in \rho(z_k \cdot y^{-1})$ or $e_2 \in \rho(x) * \rho(z_1)^{-1}$, $e_2 \in \rho(z_1) * \rho(z_2)^{-1}$, $\ldots$, $e_2 \in \rho(z_k) * \rho(y)^{-1}$ and so $\rho(x) = \rho(y)$.

Now, for every $K^*_{P_1}[x]$, $K^*_{P_1}[y] \in [P_1 : K]$, we have

$$\begin{aligned}
\varphi(K^*_{P_1}[x] \odot K^*_{P_1}[y]) &= \varphi(\{K^*_{P_1}[z] \mid z \in K^*_{P_1}[x] \cdot K^*_{P_1}[y]\} \\
&= \{\rho(z) \mid z \in K^*_{P_1}[x] \cdot K^*_{P_1}[y]\} \\
&= \rho(K^*_{P_1}[x] \cdot K^*_{P_1}[y]) \\
&= \rho(K^*_{P_1}[x]) * \rho(K^*_{P_1}[y]) \\
&= \rho(x) * \rho(y) \\
&= \varphi(K^*_{P_1}[x]) * \varphi(K^*_{P_1}[y]),
\end{aligned}$$

and $\varphi(K) = \varphi(K^*_{P_1}[e_1]) = \rho(e_1) = e_2$. Therefore, φ is a strong homomorphism.

Furthermore, if $\varphi(K^*_{P_1}[x]) = \varphi(K^*_{P_1}[y])$, then $\rho(x) = \rho(y)$ which implies that $x \cdot y^{-1} \cap K \neq \emptyset$, and so $K^*_{P_1}[x] = K^*_{P_1}[y]$. Thus, φ is a one to one mapping. ∎

4.3 Fundamental relations on weak polygroups

In this section, we consider the fundamental relation β^* defined on weak polygroups and present some results in this respect. Moreover, we define a semi-direct hyperproduct of two weak polygroups in order to obtain an extension of weak polygroups by weak polygroups.

Let $\mathcal{M} =< P, \cdot, e, ^{-1} >$ be a weak polygroup. We can define the relation β^* as the smallest equivalence relation on P such that quotient P/β^* is a group. In this case β^* is called the *fundamental equivalence relation* on P and P/β^* is called the *fundamental group*. Let $\mathcal{U}_P$ be the set of all finite products of elements of P and define the relation β on P as follows:

$$x \beta y \quad \text{if and only if } \{x, y\} \subseteq u, \text{ for some } u \in \mathcal{U}_P.$$

One can prove that the fundamental relation β^* is the transitive closure of the relation β.

The kernel of the canonical map $\varphi : P \to P/\beta^*$ is denoted by ω_P. It is easy to prove that the following statements:

(1) $\omega_P = \beta^*(e)$;
(2) $\beta^*(x)^{-1} = \beta^*(x^{-1})$, for all $x \in P$;
(3) $\beta^*(\beta^*(x)y) = \beta^*(x) \odot \beta^*(y)$, for all $x, y \in P$.

An element $x \in P$ will be called *single* if its equivalence class with respect to β^* is singleton, i.e., $\beta^*(x) = \{x\}$. We denote the set of all the single elements of P by S_P. It is straightforward to prove that for $a \in P$ and $x \in S_P$, if $x \in a \cdot y$ for some $y \in P$, then $\beta^*(a) = \{z \in P \mid zy = x\}$.

Let $\mathcal{M}_1 = <P_1, ., e_1, ^{-1}>$ and $\mathcal{M}_2 = <P_2, *, e_2, ^{-I}>$ be two weak polygroups. Then, on $P_1 \times P_2$ we can define a hyperproduct similar to the hyperproduct of polygroups as follows:

$$(x_1, y_1) \circ (x_2, y_2) = \{(a, b) \mid a \in x_1 \cdot x_2, \ b \in y_1 * y_2\}.$$

We call this the *direct product* of P_1 and P_2. It is easy to see that $P_1 \times P_2$ equipped with the usual direct product operation becomes a weak polygroup.

Lemma 4.3.1. *We have* $\mathcal{U}_{P_1 \times P_2} = \mathcal{U}_{P_1} \times \mathcal{U}_{P_2}$.

Proof. It is straightforward. ∎

Corollary 4.3.2. *Let* β_1^*, β_2^* *and* β^* *be the fundamental equivalence relations on* P_1, P_2 *and* $P_1 \times P_2$ *respectively. Then,*

$$(a, b) \ \beta^*_{P_1 \times P_2} \ (c, d) \ \Longleftrightarrow \ a \beta^*_{P_1} \ c \text{ and } b \ \beta^*_{P_2} \ d.$$

Theorem 4.3.3. *Let* β_1^*, β_2^* *and* β^* *be the fundamental equivalence relations on* P_1, P_2 *and* $P_1 \times P_2$ *respectively. Then,*

$$(P_1 \times P_2)/\beta^* \cong P_1/\beta_1^* \times P_2/\beta_2^*.$$

Proof. We consider the map

$$f : P_1/\beta_1^* \times P_2/\beta_2^* \longrightarrow (P_1 \times P_2)/\beta^*$$

with $f(\beta_1^*(x), \beta_2^*(y)) = \beta^*(x, y)$. It is easy to see that f is an isomorphism. ∎

Similar to polygroups and using the fundamental equivalence relation, we can define semidirect hyperproduct of weak polygroups.

Definition 4.3.4. Let $\mathcal{A} = <A, \cdot, e_1, ^{-1}>$ and $\mathcal{B} = <B, *, e_2, ^{-1}>$ be weak polygroups. We consider the group $AutA$ and the fundamental group B/β_B^*, let

$$\widehat{} : B/\beta_B^* \to AutA$$
$$\beta^*(b) \to \widehat{\beta^*(b)} = \widehat{b}$$

be a homomorphism of groups. Then, on $\mathcal{A} \times \mathcal{B}$ we define a hyperproduct as follows:

$$(a_1, b_1) \circ (a_2, b_2) = \{(x, y) \mid x \in a_1 \cdot \widehat{b_1}(a_2), \ y \in b_1 * b_2\}$$

and we call this the *semidirect hyperproduct of weak polygroups* $\mathcal{A}$ and $\mathcal{B}$.

Theorem 4.3.5. $\mathcal{A} \times \mathcal{B}$ *equipped with the semidirect hyperproduct is a weak polygroup.*

Proof. The proof is similar to the proof of Theorem 3.5.12. ∎

Lemma 4.3.6. *Let* $f : P_1 \longrightarrow P_2$ *be a strong homomorphism of weak polygroups and* β_1^*, β_2^* *fundamental equivalence relations on* P_1, P_2 *respectively. Then, the map* $F : P_1/\beta_1^* \longrightarrow P_2/\beta_2^*$ *defined by* $F(\beta_1^*(x)) = \beta_2^*(f(x))$ *is a homomorphism of fundamental groups.*

Proof. First, we show that F is well-defined. Suppose that $\beta_1^*(x) = \beta_1^*(y)$. Then, there exist $x_1, x_2, \ldots, x_{n+1} \in P_1$ with $x_1 = x, x_{n+1} = y$ and $u_1, \ldots, u_n \in \mathcal{U}_{P_1}$ such that $\{x_i, x_{i+1}\} \subseteq u_i$ $(i = 1, \ldots, n)$. Since f is a strong homomorphism and $u_i \in \mathcal{U}_{P_1}$, we get $f(u_i) \in \mathcal{U}_{P_2}$. Therefore, $f(x)\beta_2^* f(y)$ which implies that $\beta_2^*(f(x)) = \beta_2^*(f(y))$, and so $F(\beta_1^*(x)) = F(\beta_1^*(y))$. Thus, F is well-defined. Now, we have

$$\begin{aligned} F(\beta_1^*(x) \otimes \beta_1^*(y)) &= F(\beta_1^*(x \cdot y)) = \beta_2^*(f(x \cdot y)) \\ &= \beta_2^*(f(x) \cdot f(y)) = \beta_2^*(f(x)) \otimes \beta_2^*(f(y)) \\ &= F(\beta_1^*(x)) \otimes F(\beta_1^*(y)). \ \blacksquare \end{aligned}$$

Definition 4.3.7. Let f be a strong homomorphism from P_1 into P_2 and let β_1^*, β_2^* be the fundamental relations on P_1, P_2 respectively. Then, we define

$$\overline{ker f} = \{\beta_1^*(x) \mid x \in P_1, \ \beta_2^*(f(x)) = \omega_{P_2}\}.$$

The next corollary summarize the results of Lemma 4.3.6.

Corollary 4.3.8. *(1)* $\overline{ker f}$ *is a normal subgroup of the fundamental group*

P_1/β_1^*.

    (2) *If f is onto, then $(P_1/\beta_1^*)/\overline{kerf} \cong P_2/\beta_2^*$.*

Let $\mathcal{P}$ be the set of all weak polygroups and all strong homomorphisms. One can show that $\mathcal{P}$ is a category. We set $\mathcal{P}_{\beta*}$ the category of fundamental groups and homomorphisms of groups. Then, we have the following theorem:

Theorem 4.3.9. *Let $\mathcal{F}$ be the function from $\mathcal{P}$ into $\mathcal{P}_{\beta*}$ defined by $\mathcal{F}(P) = P/\beta^*$ and when $f : P_1 \longrightarrow P_2$ is a strong homomorphism*

$$\mathcal{F}(f) : P_1/\beta_1^* \longrightarrow P_2/\beta_2^*,$$
$$\beta_1^*(x) \longrightarrow \beta_2^*(f(x)),$$

where β_1^*, β_2^* are the fundamental relations on P_1, P_2 respectively. Then, $\mathcal{F}$ is a functor.

Proof. Clearly, $\mathcal{F}$ is well-defined. Now, we have the following:

    (1) If $P_1 \in obj\mathcal{P}$, then $P_1/\beta_1^* \in obj\mathcal{P}_{\beta*}$.

    (2) If $f : P_1 \longrightarrow P_2$ is a strong homomorphism, by Lemma 4.3.6, $\mathcal{F}(f)$ is a homomorphism of groups.

    (3) Suppose β_3^* is the fundamental relation on P_3. If $P_1 \xrightarrow{f} P_2 \xrightarrow{g} P_3$ is a sequence of strong homomorphisms in $\mathcal{P}$, then

$$\mathcal{F}(gf)(\beta_1^*(x)) = \beta_3^*(gf(x)) = \beta_3^*(g(f(x))) = \mathcal{F}(g)\beta_2^*(f(x))$$
$$= \mathcal{F}(g)(\mathcal{F}(f)(\beta_1^*(x))),$$

so $\mathcal{F}(gf) = \mathcal{F}(g)\mathcal{F}(f)$.

    (4) For every $P_1 \in obj\mathcal{P}$, we have $\mathcal{F}(1_{P_1})(\beta_1^*(x)) = \beta_1^*(1_{P_1}(x)) = \beta_1^*(x)$. Thus, $\mathcal{F}(1_{P_1}) = 1_{\mathcal{F}(P_1)}$. Therefore, $\mathcal{F}$ is a functor. ∎

4.4 Small weak polygroups

A weak polygroup is called *commutative weak polygroup* if the usual commutative axiom is valid, i.e., $x \cdot y = y \cdot x$, for all $x, y \in P$.

Proposition 4.4.1. *All weak polygroups of order 3 are commutative.*

Proof. Suppose that $< P = \{e, a, b\}, \cdot, e, ^{-1} >$ is a weak polygroup. It is enough to show that for every $x, y \in P \backslash \{e\}$ with $x \neq y$, we have $x \cdot y = y \cdot x$. We consider the following cases:

(1) $^{-1} : P \longrightarrow P$ be the identity map. Suppose that $t \in x \cdot y$. Then, $t \neq e$ and so $t = x$ or $t = y$. If $t = x$, then $x \in x \cdot y$, which implies that $y \in x^{-1} \cdot x = x \cdot x$. Hence, $x \in y \cdot x^{-1} = y \cdot x$, and so $x \cdot y \subseteq y \cdot x$. Similarly, we have $y \cdot x \subseteq x \cdot y$. Therefore, $x \cdot y = y \cdot x$.

(2) $^{-1} : P \longrightarrow P$ be the following map:

$$e \longrightarrow e, \quad a \longrightarrow b, \quad b \longrightarrow a.$$

Suppose that $t \in x \cdot y$. If $t = e$, then $y = x^{-1}$ and $e \in x \cdot x^{-1} \cap x^{-1} \cdot x$ which implies that $t = e \in y \cdot x$. Now, if $t \neq e$, then $t = x$ or $t = y$. If $t = x$, then $x \in x \cdot y$ and so $x \in x \cdot y^{-1} = x \cdot x$. Hence, $x \in x^{-1} \cdot x = y \cdot x$. Therefore, $x \cdot y \subseteq y \cdot x$. Similarly, $y \cdot x \subseteq x \cdot y$, and so $x \cdot y = y \cdot x$. ∎

Lemma 4.4.2. *Let e be a unit of a hyperstructure $(H, \cdot)$. If $x, y, z \in H$, such that $e \in \{x, y, z\}$, then $(x \cdot y) \cdot z \cap x \cdot (y \cdot z) \neq \emptyset$.*

Proof. It is straightforward. ∎

Corollary 4.4.3. *Let $(\{e, a, b\}, \cdot)$ be a commutative hyperstructure with a scalar unit e. If*

$$a \cdot (a \cdot b) \cap (a \cdot a) \cdot b \neq \emptyset \quad \text{and} \quad b \cdot (b \cdot a) \cap (b \cdot b) \cdot a \neq \emptyset,$$

then $(\{e, a, b\}, \cdot)$ is weak associative.

In [144], one can find all H_v-groups with three elements which contain a scalar unit element e. Now, we determine the set of all weak polygroups defined on a set with three elements and then we conclude that all weak polygroups with three elements are polygroups. For computing these, we program the algorithm by Mathematica 4.0.

Problem 4.4.4. Find all weak polygroups with three elements.

For $P = \{e, a, b\}$, consider the following Cayley table:

·	e	a	b
e	e	a	b
a	a	1	2
b	b	3	4

and then find all the fours, of non-empty subsets of P, in the place of the four $(1, 2, 3, 4)$ such that all conditions of Definition 4.2.1 are valid. Since by Proposition 4.4.1, every weak polygroup on the set $P = \{e, a, b\}$ is commutative, so there exist $7^3 = 343$ Cayley tables for candidate of a weak polygroup.

The following algorithm illustrate one method of doing so. We use the function *p* for finding the product of two singletons and the function *prod* for finding the product of two arbitrary sets.

```
set = {{e}, {a}, {b}};
prodtable = {};
pt = {{{e}, {a}, {b}}, {{a}, , }, {{b}, , }};
powerset = {{e}, {a}, {b}, {e, a}, {e, b}, {a, b}, {e, a, b}};
For [s0=1, s0 ≤ 7, s0++,
    pt[[2,2]]=powerset[[s0]];
    For [s1=1,s1 ≤ 7, s1++,
        pt[[2,3]]=powerset[[s1]];
        pt[[3,2]]=pt[[2,3]];
        For [s3=1,s3 ≤ 7, s3++,
            pt[[3,3]]=powerset[[s3]];
            AppendTo[prodtable,pt];
        ];
    ];
];
index[ele_]:=Module[{x=ele,y=set,t},
    For [t=1,tj=3,t++,
        If [x==y[[t]], Return [t]];
    ];
];
p[ele1_, ele2_, ind_, collect_]:=collect[[ind]][[index[ele1], index[ele2]]];
prod[set1_,    set2_,q_,    ptable_]:=Module[{f=set1,    g=set2,    h=q,
prd=ptable,u={},m,n},
    For [m=1, m ≤ Length[f],m++,
        For [n=1, n ≤ Length[g],n++,
            u=Union [u,p[{f [[m]]}, {g[[n]]}, h, prd]];
        ];
    ];
Return[u];
];
```

The following subroutine checks the weak associativity. We put the Cayley tables which are weak associative in *newpt* variable.

```
perm = {{a, a, b}, {b, b, a}};
```

```
associative = {};
For [l=1,l≤ Length[prodtable],l++,
   asctv=True;
   For [k=1, k≤ 2,k++,
      asctv =
      asctv && (   Intersection[prod[prod[{perm[[k,1]]},   {perm[[k,2]]},
l,   prodtable],              {perm[[k,3]]},   l,   prodtable],   prod[{perm[[k,1]]},
prod[{perm[[k,2]]},
         {perm[[k,3]]}, l, prodtable], l, prodtable]]=!={});
      If [!asctv, Break[]];
   ];
   If [asctv, AppendTo[associative, l]];
];
newpt=prodtable[[associative]];
```

The following subroutine checks the third condition of Definition 4.1.1.

```
funs = {{{e},{a},{b}},{{e},{b},{a}}};
newpt2 ={,};
For [o1=1,o1≤ 2,o1++,
   inv={};
   For [o2=1, o2≤ Length[newpt], o2++,
      mem3=True;
      For [o3=1, o3≤ 3, o3++,
         mem4=True;
         For [o4=1, o4≤ 3, o4++,
            mem5=True;
            For [o5=1, o5≤ Length[newpt[[o2]][[o3]][[o4]]], o5++,
               mem5=mem5                                              &&
(MemberQ[prod[{newpt[[o2]][[o3]][[o4]][[o5]]},
                  funs[[o1]][[o4]], o2, newpt], set[[o3]][[1]]])&&
                  (MemberQ[prod[funs[[o1]][[o3]],{newpt[[o2]][[o3]][[o4]][[o5]]},
                  o2, newpt], set[[o4]][[1]]]);
               If [!mem5, Break[]];
            ];
            mem4=mem4 && mem5;
            If [!mem4, Break[]];
```

```
        ];
        mem3=mem3 && mem4;
        If [!mem3, Break[]];
    ];
    If [mem3, AppendTo[inv,o2]];
  ];
newpt2[[o1]]=newpt[[inv]];
For             [u=1,u≤            Length[newpt2[[o1]]],            u++,
print[MatrixForm[newpt2[[o1]][[u]]]];];
];
```

After running this program, we obtain 15 weak polygroups as follows:

·	e	a	b
e	e	a	b
a	a	e	b
b	b	b	e,a

·	e	a	b
e	e	a	b
a	a	e	b
b	b	b	e,a,b

·	e	a	b
e	e	a	b
a	a	e,a	b
b	b	b	e,a

·	e	a	b
e	e	a	b
a	a	e,a	b
b	b	b	e,a,b

·	e	a	b
e	e	a	b
a	a	e,b	a
b	b	a	e

·	e	a	b
e	e	a	b
a	a	e,b	a
b	b	a	e,b

·	e	a	b
e	e	a	b
a	a	e,b	a,b
b	b	a,b	e,a

·	e	a	b
e	e	a	b
a	a	e,b	a,b
b	b	a,b	e,a,b

·	e	a	b
e	e	a	b
a	a	e,a,b	a
b	b	a	e

·	e	a	b
e	e	a	b
a	a	e,a,b	a
b	b	a	e,b

·	e	a	b
e	e	a	b
a	a	e,a,b	a,b
b	b	a,b	e,a

·	e	a	b
e	e	a	b
a	a	e,a,b	a,b
b	b	a,b	e,a,b

·	e	a	b
e	e	a	b
a	a	a	e,a,b
b	b	e,a,b	b

·	e	a	b
e	e	a	b
a	a	b	e
b	b	e	a

·	e	a	b
e	e	a	b
a	a	a,b	e,a,b
b	b	e,a,b	a,b

On the set P only one non identity automorphism can be defined:

$$f : \begin{cases} e \longleftrightarrow e \\ a \longleftrightarrow b \end{cases}$$

So, two weak polygroups $(P, \cdot)$ and $(P, *)$ are isomorphic if

$$a * a = f(b \cdot b), \quad a * b = f(b \cdot a), \quad b * a = f(a \cdot b) \text{ and } b * b = f(a \cdot a).$$

Thus, the fours $(a*a, a*b, b*a, b*b)$ and $(f(b \cdot b), f(b \cdot a), f(a \cdot b), f(a \cdot a))$ are isomorphic. In other words, instead of the table

·	e	a	b
e	e	a	b
a	a	1	2
b	b	3	4

take the table

·	e	a	b
e	e	a	b
a	a	4	3
b	b	2	1

and then replace a by b and b by a. The result is the isomorphic weak polygroup to the first one. Therefore, we have:

Theorem 4.4.5. *There are only 10 weak polygroups on the set $P = \{e, a, b\}$ up to isomorphism as follows:*

·	e	a	b
e	e	a	b
a	a	e	b
b	b	b	e, a

·	e	a	b
e	e	a	b
a	a	e	b
b	b	b	e, a, b

·	e	a	b
e	e	a	b
a	a	e, a	b
b	b	b	e, a

·	e	a	b
e	e	a	b
a	a	e, a	b
b	b	b	e, a, b

·	e	a	b
e	e	a	b
a	a	e, b	a, b
b	b	a, b	e, a

·	e	a	b
e	e	a	b
a	a	e, b	a, b
b	b	a, b	e, a, b

·	e	a	b
e	e	a	b
a	a	e,a,b	a,b
b	b	a,b	e,a,b

·	e	a	b
e	e	a	b
a	a	a	e,a,b
b	b	e,a,b	b

·	e	a	b
e	e	a	b
a	a	b	e
b	b	e	a

·	e	a	b
e	e	a	b
a	a	a,b	e,a,b
b	b	e,a,b	a,b

Corollary 4.4.6. *All the weak polygroups with three elements are polygroups.*

Chapter 5

Combinatorial Aspects of Polygroups

5.1 Chromatic polygroups

An important class of polygroups is derived from color schemes, a notion that extends D.G. Higman's homogeneous coherent configuration [77]. Chromatic polygroups are studied by Comer in [14; 15; 16; 18; 20] and they are obtained from certain edge colored complete graph by making multi-valued algebra out of the set of colors. The results of this section are obtained by Comer. In the definition of a color scheme presented below the relative product (or composition) of two relations is denoted by $\|$ and the inverse (or converse) of a relation is denoted by $\smile$. Suppose that $\mathcal{C}$ is a set (of colors) and ϵ is an involution of $\mathcal{C}$.

Definition 5.1.1. A *color scheme* is a system $\mathcal{V} = < V, C_x >_{x \in \mathcal{C}}$, where

(1) $\{C_x \mid x \in \mathcal{C}\}$ partitions $V^2 - I = \{(a, b) \in V^2 \mid a \neq b\}$;

(2) $C_x^\smile = C_{\epsilon(x)}$, for all $x \in \mathcal{C}$;

(3) for all $x \in \mathcal{C}$, $a \in V$ there exist $b \in V$ such that $(a, b) \in C_x$;

(4) $C_x \cap (C_y \parallel C_z) \neq \emptyset$ implies $C_x \subseteq C_y \parallel C_z$, i.e., the existence of a path colored (y, z) between two vertices joined by an edge colored x is independent of the two vertices.

Given a color scheme $\mathcal{V}$, choose a new symbol $I \notin \mathcal{C}$. (Think of I as the identity relation on V). The *algebra (color algebra)* of $\mathcal{V}$ is the system $\mathcal{M}_\mathcal{V} = < \mathcal{C} \cup \{I\}, \cdot, I, ^{-1} >$, where $x^{-1} = \epsilon(x)$ for $x \in \mathcal{C}$, $I^{-1} = I$,

$$x \cdot I = x = I \cdot x, \text{ for all } x \in \mathcal{C} \cup \{I\},$$

and for $x, y, z \in \mathcal{C}$,

$$x \cdot y = \{z \in \mathcal{C} \mid C_z \subseteq C_x \parallel C_y\} \cup \{I \mid y = x^{-1}\}.$$

It is straightforward to verify $\mathcal{M}_\mathcal{V}$ is a polygroup.

Definition 5.1.2. A polygroup is called *chromatic* if it is isomorphic to an algebra $\mathcal{M}_\mathcal{V}$, for some color scheme $\mathcal{V}$.

Example 5.1.3. A natural example of a chromatic polygroup is the system $G//H$ of all double cosets of a group G modulo a subgroup H, see Example 3.1.2 (1). In order to see that $G//H$ is chromatic, consider the color scheme $\mathcal{V} = < V, C_x >_{x \in \mathcal{C}}$, where $\mathcal{C} = (G//H) \setminus \{H\}$, $V = \{Ha \mid a \in G\}$ and for $x \in \mathcal{C}$, $C_x = \{(Ha, Hb) \mid ab^{-1} \in x\}$. It is not difficult to verify that $G//H$ is isomorphic to $\mathcal{M}_\mathcal{V}$.

Proposition 5.1.4. *Suppose that $\mathcal{A} = < A, \cdot, e, ^{-1} >$ and $\mathcal{B} = < B, \cdot, e, ^{-1} >$ are two polygroups whose elements have been renamed so that $A \cap B = \{e\}$. If $\mathcal{A} \cong \mathcal{M}_\mathcal{V}$ and $\mathcal{B} \cong \mathcal{M}_\mathcal{W}$, then the extension of $\mathcal{A}$ by $\mathcal{B}$, i.e., $\mathcal{A}[\mathcal{B}]$ is also chromatic.*

Proof. Suppose that $\mathcal{V} = < V, C_a >_{a \in A \setminus \{e\}}$. First, we introduce a family of pairwise disjoint color schemes $\{\mathcal{V}_w \mid w \in W\}$, where each $\mathcal{V}_w$ is isomorphic to $\mathcal{V}$. Assume that the vertex set of $\mathcal{V}_w$ is V_w and the isomorphism of $\mathcal{V}$ onto $\mathcal{V}_w$ sends x to x_w. We construct a scheme $\mathcal{V}[\mathcal{W}]$ in the following way. Replace each vertex w of the scheme $\mathcal{W}$ by the copy of $\mathcal{V}$ with vertex set V_w. Thus, the set of all vertices of $\mathcal{V}[\mathcal{W}]$ is just the union of all V_w's. An edge coloring using the elements of $(A \cup B) \setminus \{e\}$ as colors is introduced in the following way. For $a \in A \setminus \{e\}$ and $b \in B \setminus \{e\}$, let

$$(x_u, y_v) \in C_a \text{ if and only if } u = v \text{ and } (x, y) \in C_a \text{ (in } \mathcal{V})$$
$$(x_u, y_v) \in C_b \text{ if and only if } (u, v) \in C_b \text{ (in } \mathcal{W}).$$

It is easily seen that $\mathcal{V}[\mathcal{W}]$ is a scheme that represents $\mathcal{A}[\mathcal{B}]$. ∎

The double coset construction generalizes to the idea of a double quotient. This idea will not be needed in this section for a general polygroup but only for ordinary groups. The general notation (see Definition 3.3.10) is equivalent to the following when restricted to groups.

Definition 5.1.5. An equivalence relation θ on a group G is called a *(full) conjugation* on G if

  (1) $\theta(x)^{-1} = \theta(x^{-1})$, for all $x \in G$,
  (2) $\theta(xy) \subseteq \theta(x)\theta(y)$, for all $x, y \in G$.

The natural quotient system $G//\theta$ is a chromatic polygroup and we define

$$Q^2(\text{Group}) = \{G//\theta \mid \theta \text{ is a conjugation on some group } G\}.$$

A conjugation θ is called *special* if it satisfies

(3) $x\theta e$ implies $x = e$.

The class of all polygroups isomorphic to double quotients of groups via special conjugation is denoted by $Q_s^2(\text{Group})$.

Example 5.1.6. Some of the ways to obtain conjugations on a group G are indicated below:

(1) A congruence relation θ on G is a conjugation. $G//\theta$ is just the usual quotient group in this case.

(2) If H is a subgroup of G and we define $x\theta_H y$ if and only if x and y generate the same double coset (i.e., $HxH = HyH$), then θ_H is a conjugation on G and $G//\theta_H = G//H$.

(3) Define $x\theta y$ if and only if x and y are conjugate in the usual sense with respect to a subgroup H (i.e., there exists $h \in H$ such that $y = h^{-1}xh$). Then, θ is a special conjugation. This example generalizes as follows.

(4) Suppose that K is a group of automorphisms of G. Define $x\theta y$ if and only if $y = \sigma(x)$, for some $\sigma \in K$. This is also a special conjugation on G.

Proposition 5.1.7. *If θ is a conjugation on a group G, then $\theta(e) = H$ is a subgroup of G and $G//\theta \cong (G//H)//\psi$ for some special conjugation ψ on $G//H$.*

Proof. Suppose that $x, y \in \theta(e)$ and $z = xy^{-1}$. Then, $x = zy$. Hence, we have

$$e \in \theta(e) = \theta(x) = \theta(zy) \subseteq \theta(z)\theta(y).$$

Therefore, there exist $z' \in \theta(z)$ and $y' \in \theta(y)$ such that $e = z'y'$ which implies that $z' = y'^{-1}$. Now, we obtain

$$\theta(xy^{-1}) = \theta(z) = \theta(z') = \theta(y'^{-1}) = \theta(y^{-1}) = \theta(e).$$

Thus, $xy^{-1} \in \theta(e)$. For the second part, it is enough to define

$$HxH \; \psi \; HyH \iff x \; \theta \; y. \quad \blacksquare$$

Double quotients of groups are related to chromatic polygroups.

Theorem 5.1.8. *Every polygroup in $Q^2(\text{Group})$ is chromatic.*

Proof. Suppose that θ is a conjugation on a group G. By Proposition 5.1.7, $\theta(e) = H$ is a subgroup of G. Let $\mathcal{C} = \{\theta(g) \mid \theta(g) \neq H\}$, $\epsilon(a) = a^{-1}$ for $a \in \mathcal{C}$, $V = \{Hx \mid x \in G\}$, and for each $a \in \mathcal{C}$, set

$$C_a = \{(Hx, Hy) \in V^2 \mid xy^{-1} \in a\}.$$

It is easy to see that $\mathcal{V} = <V, C_a >_{a \in \mathcal{C}}$ is a color scheme. If $c \in a \cdot b$ (in $G//\theta$) and $(Hu, Hv) \in C_c$, there exist $r \in a$ and $s \in b$ with $uv^{-1} = rs$. Letting $z = r^{-1}u$ it follows that $(Hu, Hz) \in C_a$ and $(Hz, Hv) \in C_b$, so $(Hu, Hv) \in C_a \parallel C_b$.

Conversely, if $C_c \subseteq C_a \parallel C_b$ and $x \in c$, then $(Hx, H) \in C_c$, so there exists $z \in b$ with $xz^{-1} \in a$. Hence, $x \in a \cdot b$ and so $c \subseteq a \cdot b$ (in $G//\theta$). It now easily follows that the natural map from $G//\theta$ onto $\mathcal{M}_{\mathcal{V}}$ that sends $\theta(g)$ to $\theta(g)$ and $\theta(e)$ to I (=identity of $\mathcal{M}_{\mathcal{V}}$) is an isomorphism. ∎

The color scheme used in Theorem 5.1.8, is called the *regular color scheme* representation of $G//\theta$.

Definition 5.1.9. An automorphism of a color scheme $\mathcal{V} = <V, C_a >_{a \in \mathcal{C}}$ is a permutation σ of V such that for all $a \in \mathcal{C}$, $x, y \in V$, $(x, y) \in C_a$ if and only if $(\sigma(x), \sigma(y)) \in C_a$.

Theorem 5.1.10. *Let $\mathcal{M} = <M, \cdot, e, ^{-1}>$ be a polygroup. Then, $\mathcal{M} \in Q^2(Group)$ if and only if $\mathcal{M} \cong \mathcal{M}_{\mathcal{V}}$, for some color scheme $\mathcal{V}$ with $Aut(\mathcal{V})$ transitive on vertices.*

Proof. First note that if $\mathcal{V}$ is the regular color scheme representing $G//\theta$, then G acts transitively on the coset space V by multiplication. Now, suppose that $\mathcal{M}$ is represented by a color scheme $\mathcal{V} = <V, C_a >_{a \in \mathcal{M}}$ with $Aut(\mathcal{V}) = G$ transitive on V. Fix $x \in V$ and partition $V = \cup\{V_a \mid a \in M\}$ where $V_e = \{x\}$ and

$$V_a = \{y \in V \mid (x, y) \in C_a\} \text{ for } a \neq e.$$

Define an equivalence relation $\theta^{\mathcal{V}}$ on G by:

$$\sigma \theta^{\mathcal{V}} \tau \text{ if and only if for every } a \in M, \ \sigma(x) \in V_a \text{ if and only if } \tau(x) \in V_a.$$

The $\theta^{\mathcal{V}}$-classes correspond one to one with elements of $\mathcal{M}$. Namely, $a \in M$ corresponds to the $\theta^{\mathcal{V}}$-class $G_{xa} = \{\sigma \in G \mid \sigma^{-1}(x) \in V_a\}$ where, of course, $G_{xe} = G_x$ the stabilizer of G at x. The elements of V correspond in a natural way to cosets of G_x. Namely, for $y \in V$, $G_{xy} = \{\sigma \in G \mid y = \sigma(x)\} = G_x \tau$, where τ is any element of G_{xy}.

We obtain that $\theta^{\mathcal{V}}$ is a conjugation on G. It is easily to see that $\theta^{\mathcal{V}}$ preserves inverses. We suppose that $\sigma_0 = \sigma_2 \sigma_1$ and $\sigma_0' \theta^{\mathcal{V}} \sigma_0$ and show $\sigma_0' = \sigma_2' \sigma_1'$ for some $\sigma_1' \theta^{\mathcal{V}} \sigma_1$ and $\sigma_2' \theta^{\mathcal{V}} \sigma_2$. Suppose that $\sigma_0 \in G_{xc}$, $\sigma_1 \in G_{xa}$ and $\sigma_2 \in G_{xb}$. We have

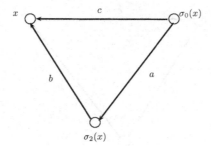

in $\mathcal{V}$. For, if $(\sigma_0(x), \sigma_2(x)) \in C_d$ for some d, applying σ_2^{-1} yields $(\sigma_1(x), x) \in C_d$ (since $\sigma_0 = \sigma_2\sigma_1$); but $(\sigma_1(x), x) \in C_a$ so $d = a$. Consequently, the colored triangle (a, b, c) is locally realizable in $\mathcal{V}$.

Now, $\sigma_0'\theta^{\mathcal{V}}\sigma_0$. Let $y = \sigma_0'(x)$. Since (a, b, c) is locally realizable and $(y, x) \in C_c$, there exists z such that $(y, z) \in C_a$ and $(z, x) \in C_b$. G vertex-transitive implies that

$$z = \tau(x), \text{ for some } \tau \in G_{xb}.$$

Choose $\mu \in G$ so that $\mu(x) = \tau^{-1}(y)$ (G vertex-transitive). Now, $(y, \tau(x)) \in C_a$ implies that $(\mu(x), x) \in C_a$, so $\mu \in G_{xa}$. Since $\tau\mu(x) = y$, both $\tau\mu$ and σ_0' belong to the coset G_{xy}. Thus, $\sigma_0' = \tau\mu v$ for some $v \in G_x$ from which it follows that $\sigma_0' = \sigma_2'\sigma_1'$, where $\sigma_1' = \mu v\theta^{\mathcal{V}}\sigma_1$ and $\sigma_2' = \tau\theta^{\mathcal{V}}\sigma_2$ as required.

In order to finish the theorem, we must show the natural bijection that sends $a \in \mathcal{M}$ to $G_{xa} \in G//\theta^{\mathcal{V}}$ is an isomorphism. First, we check that $e \in a \cdot b$ if and only if $G_x \subseteq G_{xb}G_{xa}$ (so that inverses correspond). If $e \in a \cdot b$ (so $a = b^{-1}$) and $\sigma \in G_{xa}$, then $(x, \sigma(x)) \in C_{a^{-1}} = C_b$. So, $1_{\mathcal{V}} = \sigma^{-1}\sigma \in G_{xb}G_{xa}$ and therefore $G_x \subseteq G_{xb}G_{xa}$.

Conversely, if $G_x \subseteq G_{xb}G_{xa}$, then $1_{\mathcal{V}} = \tau\sigma$ for some $\sigma \in G_{xa}$ and $\tau \in G_{xb}$. Since $(\sigma(x), x) \in C_a$, applying τ yields $(x, \tau(x)) \in C_a$. Therefore, $(\tau(x), x) \in C_{a^{-1}} \cap C_b$ so $b = a^{-1}$.

It remains to show

$$c \in a \cdot b \text{ if and only if } G_{xc} \subseteq G_{xb}G_{xa},$$

where we may assume $a, b, c \neq e$. The argument from left to right is essentially the same as used in the above. We suppose that $G_{xc} \subseteq G_{xb}G_{xa}$. By the product definition in $\mathcal{M}_{\mathcal{V}}$ it suffices to realize an (a, b, c) triangle on some $(x, y) \in C_c$. Choose σ such that $x = \sigma(y)$ (by transitivity of G). Then, $\sigma \in G_{xc} \subseteq G_{xb}G_{xa}$ so $\sigma = \mu\tau$ for some $\tau \in G_{xa}$ and $\mu \in G_{xb}$. Let $z = \tau^{-1}(x)$. Then, we have

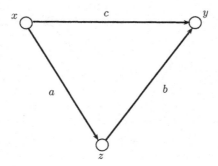

in $\mathcal{V}$. Clearly, $(x, z) \in C_a$. Suppose that $(z, y) = (\tau^{-1}(x), \tau^{-1}\mu^{-1}(x)) \in C_d$ for some d. Applying τ we obtain $(x, \mu^{-1}(x)) \in C_d$ from which $d = b$ follows. Thus, $c \in a \cdot b$ as desired. ∎

A group G of automorphisms of a graph is called *strongly transitive* on edges if, for every pair of edges (x, y) and (u, v) there exists $\sigma \in G$ such that $\sigma(x) = u$ and $\sigma(y) = v$.

Theorem 5.1.11.

> (1) $\mathcal{M}$ is isomorphic to a double coset algebra if and only if $\mathcal{M} \cong \mathcal{M}_\mathcal{V}$ for some color scheme $\mathcal{V}$ where $Aut(\mathcal{V})$ is strongly transitive on the edges of each monochrome subgraph.
>
> (2) $\mathcal{M} \in Q_s^2(Group)$ if and only if $\mathcal{M} \cong \mathcal{M}_\mathcal{V}$ for some color scheme $\mathcal{V}$ for which there exists $G \subseteq Aut(\mathcal{V})$ such that
> (i) G is vertex-transitive,
> (ii) $G_x = \{1_\mathcal{V}\}$, for all $x \in V$.

Proof. (1) We consider the regular color scheme representing $G//H$. If $(Hx, H), (Hy, H) \in C_a$, both x and y belong to the double coset a. So Hy can be obtained from Hx by right multiplication by an element of H. Thus, the regular color scheme has the desired properties. Conversely, if $Aut(\mathcal{V})$ is vertex-transitive, $\mathcal{M}_\mathcal{V} \cong G//\theta$ (Theorem 5.1.10) where $G = Aut(\mathcal{V})$ and each θ-block G_{xa} is a union of cosets $G_x\tau$ (where $x\tau^{-1} \in V_a$). G_{xa} is a single double coset since G_x is transitive on V_a.

(2) If $\mathcal{M} = G//\theta$ where $\theta(e) = \{e\}$, the vertex set in the regular color scheme representing $\mathcal{M}$ is indentifiable with G and the action of G on G is regular (i.e., $G_x = \{1\}$ for all x). Conversely, suppose that $\mathcal{V}$ representing $\mathcal{M}$ has (i) and (ii). Notice that the definition of $\theta^\mathcal{V}$ in Theorem 5.1.10 depends only on G being a transitive subgroup of $Aut(\mathcal{V})$. Thus, (i) implies

that $\mathcal{M} \cong G//\theta$ and (ii) implies that θ is special. ∎

The next three results show that various important special classes of chromatic polygroups are closed under the extension operation $\mathcal{A}[\mathcal{B}]$.

Theorem 5.1.12. *If* $\mathcal{A}, \mathcal{B} \in Q^2\,(Group)$, *then* $\mathcal{A}[\mathcal{B}] \in Q^2\,(Group)$.

Proof. Suppose that $\mathcal{A} \cong \mathcal{M}_\mathcal{V}$, $\mathcal{B} \cong \mathcal{M}_\mathcal{W}$, and $\mathcal{V}[\mathcal{W}]$ is the color scheme constructed from $\mathcal{V}$ and $\mathcal{W}$ in the proof of Proposition 5.1.4. Automorphism τ on $\mathcal{V}$ and σ on $\mathcal{W}$ induce automorphisms on $\mathcal{V}[\mathcal{W}]$ in the following way:

(1) For $\sigma \in Aut(\mathcal{W})$ define $\hat{\sigma}$ on $\mathcal{V}[\mathcal{W}]$ by $\hat{\sigma}(x_w) = x_{\sigma(w)}$, for all $w \in W$.
(2) For $\tau \in Aut(\mathcal{V})$ and $w \in W$ define $\overline{\tau}_w$ on $\mathcal{V}[\mathcal{W}]$ so that by $\overline{\tau}_w$ acts like τ on $\mathcal{V}_w$ and is the identity otherwise.

It is easy to see that the maps $\hat{\sigma}$ and $\overline{\tau}_w$ described in (1) and (2) are automorphisms of $\mathcal{V}[\mathcal{W}]$. From Theorem 5.1.10, we may assume that $Aut(\mathcal{V})$ and $Aut(\mathcal{W})$ are transitive. Using maps of type (1) and (2) it easily follows that $Aut(\mathcal{V}[\mathcal{W}])$ is transitive on vertices so Theorem 5.1.10 yields the desired conclusion. ∎

Theorem 5.1.13. *If* $\mathcal{A}, \mathcal{B} \in Q_s^2\,(Group)$, *then* $\mathcal{A}[\mathcal{B}] \in Q_s^2\,(Group)$.

Proof. Suppose that $\mathcal{A} = G_1//\theta_1$ and $\mathcal{B} = G_2//\theta_2$ where θ_1 and θ_2 are special conjugations on G_1 and G_2 respectively. Let $G = G_1 \times G_2$ and define θ on G by

$$(g_1, g_2)\theta(g_1', g_2') \iff (g_2 = g_2' = e \text{ and } g_1\theta_1g_1') \text{ or } (g_2, g_2' \neq e \text{ and } g_2\theta_2g_2').$$

Notice that the θ-classes are $\theta(g, e) = \{(h, e) \mid h\theta_1 g\}$ and for $h \neq e$, $\theta(e, h) = \{(g, h') \mid h'\theta_2 h\}$. In order to show θ is a special conjugation, conditions (1), (2) and (3) need to be checked. First, (3) holds because θ_1 special implies that $\theta(e, e) = \{(e, e)\}$. Also (1), $\theta(g, h)^{-1} = (\theta(g, h))^{-1}$, holds since θ_1 and θ_2 have similar properties. It remains to check

(2) $\theta((g_1, h_1)(g_2, h_2)) \subseteq \theta(g_1, h_1)\theta(g_2, h_2)$.

Suppose that $(g, h) \in \theta(g_1 g_2, h_1 h_2) = \theta((g_1, h_1)(g_2, h_2))$. The definition of θ gives two cases:

Case (1) $h = h_1 h_2 = e$ and $g\theta_1(g_1 g_2)$.
  Since θ_1 is a conjugation, $g = g_1' g_2'$ for some $g_1'\theta_1 g_1$ and $g_2'\theta_1 g_2$. Then, $(g, e) = (g_1', h_1)(g_2', h_2)$, so it is suffices to show that $(g_i', h_i)\theta(g_i, h_i)$ for $i = 1, 2$. The conclusion follows from $g_i'\theta_1 g_i$ when $h_1 = h_2 = e$ while it follows from $h_i\theta_2 h_i$ if $h_1, h_2 \neq e$.

Case (2) $h, h_1 h_2 \neq e$ and $h \theta_2 h_1 h_2$.

Since θ_2 is a conjugation, $h = h'_1 h'_2$ for some $h'_1 \theta_2 h_1$ and $h'_2 \theta_2 h_2$. This yields $(g, h'_1) \theta(g_1, h_1)$ and $(e, h'_2) \theta(g_2, h_2)$ whenever $h'_1, h'_2 \neq e$; so

$$(g, h) = (g, h'_1)(e, h'_2) \in \theta(g_1, h_1) \theta(g_2, h_2).$$

On the other hand, suppose that one of the h'_1, h'_2 is e, say $h'_1 = e$ and $h'_2 = h \neq e$. Then, $(g_1, e) \theta(g_1, h_1)$ and $(g_1^{-1} g, h) \theta(g_2, h)$ since $h \neq e$ from which it follows that (g, h) belongs to $\theta(g_1, h_1) \theta(g_2, h_2)$.

Thus, θ is a special conjugation on the group $G_1 \times G_2$. A bijection F between the elements of $\mathcal{A}[\mathcal{B}]$ and $G_1 \times G_2 // \theta$ is defined in the following way. Let $F(e) = \theta(e, e)$ and for $a = \theta_1(g_1) \neq e$ in $A \setminus \{e\}$, let

$$F(a) = \theta(g_1, e),$$

and for $b = \theta_2(g_2) \neq e$ in $B \setminus \{e\}$, let

$$F(b) = \theta(e, g_2).$$

From the description of the θ-classes above it is clear that the map F from $\mathcal{A}[\mathcal{B}]$ to $G_1 \times G_2 // \theta$ is a bijection. By properties (1) and (2) of θ the inverse and identity elements correspond. Computation, as in the proof of (2), show that F preserves products in case at least one factor belong to $\mathcal{A}$. When both factors belong to $\mathcal{B}$ there are two cases. First, if $\theta_2(g_2), \theta_2(g'_2) \in B \setminus \{e\}$ and $\theta_2(g_2) \neq (\theta_2(g'_2))^{-1}$, then

$$\begin{aligned}
F(\theta_2(g_2)\theta_2(g'_2)) &= F(\{\theta_2(g) \mid e \neq g \in \theta_2(g_2)\theta_2(g'_2)\}) \\
&= \{\theta(e, g) \mid g \in \theta_2(g_2)\theta_2(g'_2)\} \\
&= \{(h, g) \mid g \in \theta_2(g_2)\theta_2(g'_2)\} \\
&= \theta(e, g_2)\theta(e, g'_2) \\
&= F(\theta_2(g_2))F(\theta_2(g'_2)).
\end{aligned}$$

Finally, suppose that $\theta_2(g_2), \theta_2(g'_2) \in B - \{e\}$ and $\theta_2(g'_2) = (\theta_2(g_2))^{-1}$. Then, $e \in \theta_{(}g_2)\theta_2(g'_2)$ so

$$\begin{aligned}
F(\theta_2(g_2)\theta_2(g'_2)) &= F(\{\theta_2(g) \mid g \in \theta_2(g_2)\theta_2(g'_2)\} \cup A) \\
&= \{\theta(e, g) \mid g \in \theta_2(g_2)\theta_2(g'_2)\} \cup (G_1 \times \{e\}) \\
&= \{(h, g) \mid g \in \theta_2(g_2)\theta_2(g'_2)\} \text{ since } e \in \theta_{(}g_2)\theta_2(g'_2) \\
&= \theta(e, g_2)\theta(e, g'_2) \\
&= F(\theta_2(g_2))F(\theta_2(g'_2)).
\end{aligned}$$

Now, the theorem follows from the fact F is an isomorphism. ∎

The next result shows the class of double coset algebras is closed under the extension construction.

Theorem 5.1.14. *If $\mathcal{A} \cong G_1//H_1$ and $\mathcal{B} \cong G_2//H_2$, then there exist groups $\widehat{G}$ and $\widehat{H}$ with $\widehat{H} \subseteq \widehat{G}$ such that $\mathcal{A}[\mathcal{B}] \cong \widehat{G}//\widehat{H}$.*

Proof. Let $X = G_2/H_2 = \{H_2 g \mid g \in G_2\}$ and let $\widehat{G} = G_1^X \times_\varphi G_2$, the semidirect product of G_1^X by G_2, where φ, mapping G_2 into $(Aut(G_1^X))$, is the homomorphism given by $\varphi_g(f) = g^* f$, for all $f \in G_1^X$.

$$X \xrightarrow{\;\;g^*\;\;} X$$
$$\varphi_g(f) \searrow \;\;\swarrow f$$
$$G_1$$

i.e., $g \in G_2$ induces $g^* : X \longrightarrow X$ by right multiplication, so $\varphi_g(f)(x) = f(xg)$ for $g \in G_2$, $f \in G_1^X$ and $x \in X$. Also, let $\widehat{H} = (H_1 \times G_1^{X-\{H_2\}}) \times_\varphi H_2$, where

$$\varphi : H_2 \longrightarrow Aut(H_1 \times G_1^{X-\{H_2\}})$$

is defined, as above, by $\varphi_g(f) = g^* f$.

Notice that $\widehat{H} = \overline{H} \times_\varphi H_2$, where $\overline{H} = \{f \in G_1^X \mid f(H_2) \in H_1\}$. Clearly, $\widehat{H}$ is a subgroup of $\widehat{G}$, so it remains to show that $\mathcal{A}[\mathcal{B}] \cong \widehat{G}//\widehat{H}$.

First, we identify $\mathcal{A}$ with part of $\widehat{G}//\widehat{H}$. For $g \in G_1$ let $\widehat{g} = (f, 1)$ where 1 is the identity element of G_2 and let $f \in G_1^X$ is defined by

$$f(H_2 x) = \begin{cases} g & \text{if } H_2 x = H_2 \\ e & \text{if } H_2 x \neq H_2. \end{cases}$$

Then,

$$\widehat{H}\widehat{g}\widehat{H} = (\overline{H} \times_\varphi H_2)(f, 1)(\overline{H} \times_\varphi H_2) = (\overline{H} f \times_\varphi H_2)(\overline{H} \times_\varphi H_2) = (\overline{H} f \overline{H}) \times_\varphi H_2,$$

where the second equality holds because, for $h \in H_2$, φ_h fixes the "H_2-coordinate" of f. Thus,

(1) $\quad (G_1^X \times_\varphi H_2)//\widehat{H} \cong G_1//H_1$.

Now, we consider the elements in $G_2//H_2$ ($\cong \mathcal{B}$). For $g \in G_2$ let $\overline{g} = (E, g)$ where E is the identity element of G_1^X. Then,

$$\widehat{H}\overline{g} = (\overline{H} \times_\varphi H_2)\overline{g} = \overline{H} \times_\varphi H_2 g$$

and, for $g \notin H_2$,

$$\widehat{H}\overline{g}\widehat{H} = (\overline{H} \times_\varphi H_2 g)(\overline{H} \times_\varphi H_2) = G_1^X \times_\varphi (H_2 g H_2)$$

since, for $f_1, f_2 \in \overline{H}$, $(f_1 \cdot f_2^{hg})(x) = f_1(x) f_2(xhg)$ will produce any element of G_1^X. In order to see this observe that $g \notin H_2$ means g^* is a permutation of X that moves H_2, so for all $x \in X$, either $f_1(x)$ or $f_2(xhg)$ can be any element of G_1. Thus,

(2) $\widehat{G} = \bigcup \{ G_1^X \times_\varphi b \mid b \in G_2 // H_2 \}$.

A one to one correspondence between the nonidentity elements of $(G_1 // H_1)[G_2 // H_2]$ and $\widehat{G} // \widehat{H}$ is introduced as follows: to every nonidentity element $a \in G_1 // H_1$ assign the element $\overline{a} \times_\varphi H_2$ where $\overline{a} = \{ f \in G_1^X \mid f(H_2) \in a \}$ and to every nonidentity element $b \in G_2 // H_2$ correlate $G_1^X \times_\varphi b$. In view of (1), to show this correspondence is an isomorphism it is enough to check:

(3) $(\widehat{H} \widehat{g_1} \widehat{H})(\widehat{H} \overline{g_2} \widehat{H}) = \widehat{H} \overline{g_2} \widehat{H}$, for $g_1 \in G_1, g_2 \in G_2 \setminus H_2$,

(4) $(\widehat{H} \overline{g_2} \widehat{H})(\widehat{H} \widehat{g_1} \widehat{H}) = \widehat{H} \overline{g_2} \widehat{H}$, for $g_1 \in G_1, g_2 \in G_2 \setminus H_2$,

(5) $(\widehat{H} \overline{g_1} \widehat{H})(\widehat{H} \overline{g_2} \widehat{H}) = G_1^X \times_\varphi (H_2 g_1 H_2 g_2 H_2)$, for $g_1, g_2 \in G_2 \setminus H_2$.

To establish (3), let $\widehat{g_1} = (g_1', 1)$. Then,

$$(\widehat{H} \widehat{g_1} \widehat{H})(\widehat{H} \overline{g_2} \widehat{H}) = ((\overline{H} g_1' \overline{H}) \times_\varphi H_2)(G_1^X \times_\varphi H_2 g_2 H_2)$$
$$= G_1^X \times_\varphi H_2 g_2 H_2$$
$$= \widehat{H} \overline{g_2} \widehat{H}$$

since, for $h \in H_2$, φ_h fixes the "H_2-coordinate" and permutes all others.

The verification of (4) is easier:

$$(\widehat{H} \overline{g_2} \widehat{H})(\widehat{H} \widehat{g_1} \widehat{H}) = (G_1^X \times_\varphi H_2 g_2 H_2)(\overline{H} g_1' \overline{H} \times_\varphi H_2)$$
$$= G_1^X \times_\varphi H_2 g_2 H_2.$$

Finally, we check (5):

$$(\widehat{H} \overline{g_1} \widehat{H})(\widehat{H} \overline{g_2} \widehat{H}) = (G_1^X \times_\varphi H_2 g_1 H_2)(G_1^X \times_\varphi H_2 g_2 H_2)$$
$$= G_1^X \times_\varphi (H_2 g_1 H_2)(H_2 g_2 H_2).$$

Now, it follows that $\widehat{G} // \widehat{H} \cong \mathcal{A}[\mathcal{B}]$ as desired. ∎

Now, we consider the converse of the properties established in Theorems 5.1.4, 5.1.12, 5.1.13 and 5.1.14. The basic idea for establishing the converses is illustrated by the proof of the following result.

Theorem 5.1.15. *If $\mathcal{A}[\mathcal{B}]$ is chromatic, then both $\mathcal{A}$ and $\mathcal{B}$ are chromatic.*

Proof. Let $\mathcal{A}[\mathcal{B}] \cong \mathcal{M}_\mathcal{W}$ for some color scheme $\mathcal{W} = \langle W, C_x \rangle_{x \in \mathcal{C}}$. Recall that $\mathcal{C} = (A \cup B) \setminus \{e\}$. We define a relation $\approx$ on W by

$w \approx w'$ if and only if $w = w'$ or $(w, w') \in C_a$ for some $a \in A \setminus \{e\}$.

It is easy to see that $\approx$ is an equivalence relation on W and each $\approx$-block, say $[p] = \{ w \mid w \approx p \}$ for a fixed $p \in W$, inherits the structure of a color scheme from $\mathcal{W}$. The color algebra of this scheme is exactly $\mathcal{A}$, so $\mathcal{A}$ is chromatic.

In order to treat $\mathcal{B}$ we form a new scheme $\mathcal{W}/\approx$ on the set $\{[w] \mid w \in W\}$ using the elements of $B - \{e\}$ as colors. For distinct vertices $[v]$ and $[w]$ set

$$([v], [w]) \in C_b \text{ if and only if } (v, w) \in C_b \text{ (in } \mathcal{W}).$$

Since $a_1 b a_2 = b$ holds in $\mathcal{A}[\mathcal{B}]$ for $a_1, a_2 \in A$ and $b \in B$, it follows that the assignment of a color to the edge $([v], [w])$ is independent of the $\approx$-representations. It is not difficult to check that $\mathcal{M}_\mathcal{W}/\approx \cong \mathcal{B}$ as desired. ∎

Using the idea above an analysis of the proofs of Theorems 5.1.12, 5.1.13 and 5.1.14 give a hint of how to construct their converses. We leave the details to the reader.

Theorem 5.1.16. *If $\mathcal{A}[\mathcal{B}]$ is a double coset algebra (in $Q^2(Group)$, $Q_s^2(Group)$), then both $\mathcal{A}$ and $\mathcal{B}$ are double coset algebras (in $Q^2(Group)$, $Q_s^2(Group)$, respectively).*

Suppose that $\mathcal{V}_0 = <V_0, C_\alpha>_{\alpha \in A - \{e\}}$ and $\mathcal{V}_1 = <V_1, C_\beta>_{\beta \in B - \{e\}}$ are color schemes. It is convenient to assume the identity relation occurs among the colors, i.e., $C_e = I_{V_0}$ (in $\mathcal{V}_0$) and $C_e = I_{V_1}$ (in $\mathcal{V}_1$). Consider the product scheme

$$\mathcal{V}_0 \times \mathcal{V}_1 = <V_0 \times V_1, C_{(\alpha,\beta)}>_{(\alpha,\beta) \in A \times B - \{(e,e)\}}$$

where $C_{(\alpha,\beta)} = \{((a,i),(b,j)) \in (V_0 \times V_1)^2 \mid (a,b) \in C_\alpha \text{ and } (i,j) \in C_\beta\}$.

Theorem 5.1.17. $\mathcal{V}_0 \times \mathcal{V}_1$ *is a color scheme and* $\mathcal{M}_{\mathcal{V}_0 \times \mathcal{V}_1} = \mathcal{M}_{\mathcal{V}_0} \times \mathcal{M}_{\mathcal{V}_1}$.

Proof. Most of the color scheme properties are trivial. We show (4). Suppose that $((a,i),(b,j))$ belongs to $C_{(\alpha_0,\beta_0)} \cap (C_{(\alpha_1,\beta_1)} \parallel C_{(\alpha_2,\beta_2)})$. If $\alpha_k = e$ or $\beta_k = e$ for some $k \in \{0,1,2\}$, the condition above degenerates and the conclusion is clear. Assume that all α's and β's are $\neq e$. Then, $(a,b) \in C_{\alpha_0}$ and $(i,j) \in C_{\beta_0}$ and there exists (c,k) such that

$$(a,c) \in C_{\alpha_1}, \ (i,k) \in C_{\beta_1}, \ (c,b) \in C_{\alpha_2} \text{ and } (k,j) \in C_{\beta_2}.$$

Then, $C_{\alpha_0} \cap (C_{\alpha_1} \parallel C_{\alpha_2}) \neq \emptyset$ and $C_{\beta_0} \cap (C_{\beta_1} \parallel C_{\beta_2}) \neq \emptyset$, so $C_{\alpha_0} \subseteq C_{\alpha_1} \parallel C_{\alpha_2}$ and $C_{\beta_0} \subseteq C_{\beta_1} \parallel C_{\beta_2}$. It follows that $C_{(\alpha_0,\beta_0)} \subseteq C_{(\alpha_1,\beta_1)} \parallel C_{(\alpha_2,\beta_2)}$. In order to see that $\mathcal{M}_{\mathcal{V}_0 \times \mathcal{V}_1} = \mathcal{M}_{\mathcal{V}_0} \times \mathcal{M}_{\mathcal{V}_1}$, notice that allowing $\alpha = e$ and $\beta = e$ simplifies the definition of $*$ in $\mathcal{M}_{\mathcal{V}_0 \times \mathcal{V}_1}$. Namely,

$$(\alpha_0, \beta_0) * (\alpha_1, \beta_1) = \{(\alpha_2, \beta_2) \mid C_{(\alpha_2,\beta_2)} \cap (C_{(\alpha_0,\beta_0)} \parallel C_{(\alpha_1,\beta_1)}) \neq \emptyset\}.$$

As in the proof of (4) above, the product

$$(\alpha_2, \beta_2) \in (\alpha_0, \beta_0) * (\alpha_1, \beta_1)$$

is equivalent to products $\alpha_2 \in \alpha_0 \cdot \alpha_1$ and $\beta_2 \in \beta_0 \cdot \beta_1$ on the factors. This, in turn, is equivalent to the product in $\mathcal{M}_{\mathcal{V}_0} \times \mathcal{M}_{\mathcal{V}_1}$. ∎

What other classes are closed under direct product? For color scheme $\mathcal{V}_0$ and $\mathcal{V}_1$ it is not hard to see that

Lemma 5.1.18. $Aut(\mathcal{V}_0) \times Aut(\mathcal{V}_1) \cong Aut(\mathcal{V}_0 \times \mathcal{V}_1)$.

Proof. Consider the map that sends $(\sigma_0, \sigma_1) \in Aut(\mathcal{V}_0) \times Aut(\mathcal{V}_1)$ to the automorphism $[\sigma_0, \sigma_1]$ defined as $[\sigma_0, \sigma_1](a, i) = (\sigma_0(a), \sigma_1(i))$ for all $(a, i) \in V_0 \times V_1$. ∎

Theorem 5.1.19. *The following classes are closed under direct product:*

  (1) Q^2*(Group)*;
  (2) Q_s^2*(Group)*;
  (3) *double coset algebras.*

Proof. (1) By Theorem 5.1.10, $\mathcal{M} \in Q^2$(Group) if and only if $\mathcal{M} \cong \mathcal{M}_{\mathcal{V}}$ for some color scheme $\mathcal{V}$ with $Aut(\mathcal{V})$ transitive on vertices. Given $\mathcal{V}_0$, $\mathcal{V}_1$ both with such automorphism groups Lemma 5.1.18 shows that $Aut(\mathcal{V}_0 \times \mathcal{V}_1)$ is also transitive on vertices. By Theorem 5.1.17, $\mathcal{M}_{\mathcal{V}_0} \times \mathcal{M}_{\mathcal{V}_1} \in Q^2$(Group). The proof of (2) and (3) is similar using Theorem 5.1.11 items (1) and (2). ∎

We mention one application of direct product. We associate a polygroup $\mathcal{M}(L)$ with each modular lattice $L = < L, \vee, \wedge >$ with a minimum element e. Namely, $\mathcal{M}(L) = < L, \cdot, \ ^{-1}, e >$ where $x^{-1} = x$ and

$$x \cdot y = \{z \in L \mid x \vee z = y \vee z = x \vee y\}.$$

It is not hard to check that $\mathcal{M}(L_1 \times L_2) = \mathcal{M}(L_1) \times \mathcal{M}(L_2)$. Theorem 5.1.17 implies that whenever the $\mathcal{M}(L)$ construction associates a chromatic polygroup to latices L_1 and L_2, the product also gives rise to a chromatic polygroup. A similar conclusion holds for the classes listed in Theorem 5.1.19.

Recall that the ordered sum $L_0 \oplus L_1$ of two bounded lattices is the lattice obtained by identifying the minimum element of L_1 with the maximal element L_0.

Corollary 5.1.20. *For bounded modular lattices L_0 and L_1, $\mathcal{M}(L_0 \oplus L_1) = \mathcal{M}(L_0)[\mathcal{M}(L_1)]$. Thus, $L_0 \oplus L_1$ yields a chromatic polygroup whenever L_0 and L_1 do.*

Composition series play an important role in the study of groups. Polygroups also exhibit similar series. The *core* of a polygroup $\mathcal{M}$, written

$Core(\mathcal{M})$, is the subpolygroup generated by $\cup\{a \cdot a^{-1} \mid a \in M\}$. For a subpolygroup $\mathcal{N}$ of $\mathcal{M}$, we introduce a conjugation $\Theta_{\mathcal{N}}$ on $\mathcal{M}$ by $a\Theta_{\mathcal{N}}b$ if and only if $b \in NaN$. Then, $\mathcal{M}//\Theta_{\mathcal{N}}$ is a polygroup.

Definition 5.1.21. An *ultragroup* is a polygroup $\mathcal{M}$ for which exists a chain of subpolygroups

$$\{e\} = \mathcal{N}_k \subseteq \mathcal{N}_{k-1} \subseteq \ldots \subseteq \mathcal{N}_0 = \mathcal{M},$$

where $\mathcal{N}_i//\Theta_{\mathcal{N}_{i-1}}$ is a group for all $i < k$. The groups $\mathcal{N}_i//\Theta_{\mathcal{N}_{i-1}}$ are called the *factors of the series*. Note that $Core(\mathcal{A}[\mathcal{B}]) = A \cup Core(\mathcal{B})$ so if $\mathcal{B}$ is a group, $Core(\mathcal{A}[\mathcal{B}]) = A$.

Proposition 5.1.22. *Given groups* $G_0, G_1, \ldots, G_{k+1}$ *and* $0 < i < k$, *let* $\mathcal{N}_i$ *denote the extension* $(\ldots (G_0[G_1])\ldots)[G_{i-1}]$. *Then,*

(1) $\mathcal{N}_k$ *is an ultragroup with factors* $G_{k-1}, \ldots, G_0$;
(2) $\mathcal{N}_k$ *is a double coset algebra.*

Proof. (1) Since $\mathcal{N}_i = \mathcal{N}_{i-1}[G_{i-1}]$ and $Core(\mathcal{N}_i) = \mathcal{N}_{i-1}$, $\mathcal{N}_k \supset \mathcal{N}_{k-1} \supset \ldots \supset \mathcal{N}_1 \supset \{e\}$ gives the lower ultra-series for $\mathcal{N}_k$. Moreover, note that for $a, b \in N_i$,

$$a \, \Theta_{\mathcal{N}_{i-1}} \, b \text{ if and only if } a, b \in N_{i-1} \text{ or } a = b.$$

Then, $\mathcal{N}_i//\Theta_{\mathcal{N}_{i-1}} = \mathcal{N}_i//\mathcal{N}_{i-1} = G_{i-1}$.
(2) is obvious. ∎

5.2 Polygroups derived from cogroups

Cogroups were introduced by Eaton [70] in an attempt to axiomatize D-hypergroups, i.e., systems obtainable from groups by right coset decompositions with respect to, not necessarily normal, subgroups. Eaton's axioms apply only to finite systems. Utimi [141] formulated a general notion and gave an example of a cogroup not isomorphic to a D-hypergroup. The definition of a cogroup given by Comer [14] is equivalent to the one formulated by Utuni [141]. The cardinality axiom assumed by both Utumi and Eaton does not play a role in Comer's development. The apparently weaker notion of weak cogroup is obtained by removing this assumption. The main reference for this section is [14].

Definition 5.2.1. A *weak cogroup* is a system $< A, \cdot, \, ^{-1}, e >$, where $e \in A$; $x \cdot y$ is a non-empty subset of A for $x, y \in A$; x^{-1} is a non-empty subset of A for $x \in A$; and the following axioms hold for all $x, y, z \in A$;

(1) $(x \cdot y) \cdot z = x \cdot (y \cdot z)$,

(2) $e \cdot x = x$,

(3) $y \in x^{-1} \iff e \in x \cdot y$,

(4) $x \in y \cdot z \implies y \in x \cdot z^{-1}$ and $z \in y^{-1} \cdot x$,

(5) $x \cdot y \cap z \cdot y \neq \emptyset \implies x \in z \cdot e$.

A weak cogroup is called a *cogroup* if, in addition, it satisfies the axiom

(6) $|x \cdot y| = |x \cdot z|$, for all $x, y, z \in A$.

If H is a subgroup of a group G, the system $G/H =< \{Hg \mid g \in G\}, \cdot, {}^{-1}, H >$ of all right cosets becomes a cogroup using the operation $(Hg) \cdot (Hk) = \{Hghk \mid h \in H\}$ and $(Hg)^{-1} = \{Hg^{-1}h \mid h \in H\}$. The system G/H is known as a *D-hypergroup*.

Elements x and y in a weak cogroup are called *e-conjugates*, in symbols $x \approx y$ if and only if $x \in y \cdot e$. It is easy to see that $\approx$ is an equivalence relation on A, the $\approx$-class of e is $\{e\}$, and

$$x \in y \cdot e \Leftrightarrow x \cdot e = y \cdot e \Leftrightarrow x \cdot z = y \cdot z \text{ for all } z.$$

The product $x \cdot e$ is the $\approx$-class that contains x.

The canonical example of a cogroup is G/H while that of a polygroup is $G//H$. Every element in $G//H$ is a $\approx$-class of G/H. This suggests a way to construct polygroups from arbitrary weak cogroups.

Lemma 5.2.2. *Suppose that A is a weak cogroup and $\approx$ is the relation of e-conjugation. Then,*

(1) $(a \cdot e) \cdot (b \cdot e) = (a \cdot b) \cdot e = \{c \cdot e \mid c \in a \cdot b\}$ and $(a \cdot e)^{-1} = a^{-1} \cdot e = a^{-1}$.

(2) The system $A/ \approx$ of all $\approx$-classes, with operations inherited from A, is a polygroup.

Proof. It is straightforward. As an example consider $a^{-1} \cdot e \subseteq a^{-1}$. Suppose that $b \in c \cdot e$, for some $c \in a^{-1}$. Then, $e \in a \cdot c$, which implies that $e \in c \cdot a = b \cdot a$. Hence, $e \in a \cdot b$ and so $b \in a^{-1}$. ∎

The system $A/ \approx$ is called the *polygroup derived from A*.

The following definition and lemma are essentially due to Utumi [141]. For an equivalence relation θ on a weak cogroup $< A, \cdot, {}^{-1}, e >$, let

$$A^{*\theta} =< A, *, {}^{-1}, e^* >$$

where $e^* = e$, $x * y = \theta(x) \cdot y$ and $x^{-1} = \theta(x)^{-1}$, i.e., $x^{-1} = \cup \{y^{-1} \mid y \in \theta(x)\}$. $A^{*\theta}$ is called a *scalar partition hypergroupoid* with respect to θ. We refer to the structure as A^* whenever θ is understood.

Lemma 5.2.3. *Suppose that A is a weak cogroup and θ is an equivalence relation on A with $x \cdot e \subseteq \theta(x)$, for all x. Then, A^* is a weak cogroup if and only if*

(1) $\theta(e) = \{e\}$,
(2) $\theta(x^{-1}) = (\theta(x))^{-1}$,
(3) $\theta(\theta(x)y) = \theta(x)\theta(y)$.

Moreover, if A is a cogroup, so is A^*.

Proof. It is straightforward. ■

The condition $x \cdot e \subseteq \theta(x)$ in Lemma 5.2.3 is only needed to establish the implication $\Rightarrow$.

For either a weak cogroup A or a polygroup $\mathcal{M}$, an equivalence relation θ on A that satisfies conditions (1),(2), (3) in Lemma 5.2.3 and has $x \cdot e \subseteq \theta(x)$ for all $x \in A$ is called an *Utumi partition*. The condition $x \cdot e \subseteq \theta(x)$ for all $x \in A$, is redundant when $\mathcal{A}$ is a polygroup. These partitions are closely related to special conjugations.

In Definition 3.3.10, we introduced the notion of conjugation and in Lemma 3.3.11, we gave a necessary and sufficient condition for an equivalence relation be conjugation. Notice that θ is a special conjugation if and only if θ is a Utumi partition.

Corollary 5.2.4. *If θ is a special conjugation on a group G, then $G^{*\theta}$ is a cogroup and $G^{*\theta}/\approx = G//\theta$.*

Proof. The equality of quotients follows because $x \approx y$ if and only if $x \in y^*e = \theta(y)$ if and only if $\theta(x) = \theta(y)$. ■

The corollary will be generalized to conjugation.

Suppose that H is a subgroup of a group G and π is a Utumi partition on G/H. Let

$$G[H, \pi] = (G/H)^{*\pi}/\approx .$$

Also, we define $\overline{\pi}$ on G^2 by

$$g_1 \overline{\pi} g_2 \Longleftrightarrow (\pi(Hg_1))H = (\pi(Hg_2))H.$$

For readability, $(\pi(A))B$ is written as $\pi(A)B$.

Lemmas 5.2.2 and 5.2.3 show that $G[H, \pi]$ is a polygroup. Also, in $(G/H)^*$, $Hg_1 \approx Hg_2$ if and only if $\pi(Hg_1)H = \pi(Hg_2)H$. Thus, $\pi(Hg)H$ is the $\approx$-class of Hg. Also, $\pi(Hg)H = \pi(Hg)$ in $(G/H)^*$ since a Utumi partition satisfies $x \cdot e \subseteq \pi(x)$. Thus, $\approx$ coincides with π.

Lemma 5.2.5. $\overline{\pi}$ *is a conjugation on* G *and* $G[H, \pi] \cong G//\overline{\pi}$.

Proof. Clearly, $\overline{\pi}$ is an equivalence relation on G. In order to check condition (1) for a conjugation, suppose that $x\overline{\pi}y$. Then, $Hx \approx Hy$ in $(G/H)^*$. Since Hx^{-1} is an inverse of Hx, e-conjugate elements have the same inverses, and every two inverses of an element are e-conjugate, $Hx^{-1} \approx Hy^{-1}$. Thus, $x^{-1}\overline{\pi}y^{-1}$ and (1) holds.

In order to check the condition (2) for $\overline{\pi}$ to be a conjugation, assume that $x'\overline{\pi}x$ and $x = y \cdot z$. Then, $Hx' \in \pi(Hx)$ and $Hx \in (Hy)(Hz)$; so, in G/H,

$$\begin{aligned} Hx' \in \pi(Hx) &\subseteq \pi(Hy)(Hz)) \\ &\subseteq \pi(\pi(Hy)Hz) \\ &= \pi(Hy)\pi(Hz) \end{aligned}$$

by the Utumi properties. Thus, $Hx' \subseteq (Hy'')(Hz'')$ for some $Hy'' \in \pi(Hy)$ and $Hz'' \in \pi(Hy)$. Then, $x' = y' \cdot z'$ for some $y' \in Hy''$ and $z' \in Hz''$. Since $\pi(Hy') = \pi(Hy'') = \pi(Hy)$, $y'\overline{\pi}y$. Similarly, $z'\overline{\pi}z$. Thus, (2) holds and $\overline{\pi}$ is a conjugation on G.

The correspondence that sends $\overline{\pi}$ to $Hg/\approx= \pi(Hg)H = \pi(Hg)$ is clearly a bijection between polygroups $G//\overline{\pi}$ and $G[H, \pi]$. The identity elements correspond as well inverses since $(\overline{\pi}(x))^{-1} = \pi(x^{-1})$ and $(\pi(Hx))^{-1} = \pi(Hx^{-1})$. In order to see that the correspondence is an isomorphism observed that $\pi(Hx) \in \pi(Hy) \cdot \pi(Hz)$ is equivalent to $\overline{\pi}(x) \in \overline{\pi}(y) \cdot \overline{\pi}(z)$.∎

The lemma above shows that polygroups derived from Utumi partitions on D-hypergroups are double quotients of groups. The next result establishes the converse. Namely, every double quotient of a group is derivable from a D-hypergroup with a Utumi partition.

Theorem 5.2.6. *For any* $G//\theta \in Q^2$ *(Group) there exist a subgroup* H *of* G *and a Utumi partitions* π *on* G/H *such that* $G//\theta \cong G[H, \pi]$.

Proof. Given G and θ, $H = \theta(e)$ is a subgroup of G. Define π on $G?H$ by $(Hg_1)\pi(Hg_2)$ if and only if $g_1\theta g_2$. Since θ is a conjugation on G, $Hg_1 = Hg_2$ implies that $g_1\theta g_2$, from which it follows that π is well defined (i.e., it factors through the quotient mod H). Clearly, π is an equivalence relation on G/H and, since a conjugation θ has the property $g\theta gh$ for all $h \in H$, $(Hg)H \subseteq \pi(Hg)$. We claim that

  (1) π is a Utumi partition on G/H.

Since $H = \theta(e)$, $(Hg)\pi H$ easily implies that $Hg = H$. It remains to check (2) and (3) of Lemma 5.2.3. First, we show that

$$(\pi(Hg))^{-1} = \pi(Hg^{-1}).$$

Suppose that $Hg_0 \in (\pi(Hg))^{-1}$. Then, $H \in Hg_0Hg'$ for some $g'\theta g$ which yields that $g'^{-1}\theta g_0$. Since θ is a conjugation. $g'^{-1}\theta g^{-1}$, from which we obtain $Hg_0 \in \pi(Hg^{-1})$. The converse is similar. Now, we consider

$$\pi(\pi(Hg_1))Hg_2) = \pi(Hg_1)\pi(Hg_2).$$

In order to establish $\subseteq$ assume that $(Hg)\pi(Hg') \in (\pi(Hg_1))(Hg_2)$ for some $g'\theta g$. It follows that $g' = hg_1'h'g_2 = g''g_2$ for some $h, h' \in H$ and $g_1'\theta g_1$, where $g'' = hg_1'h'$. Since θ is a conjugation, it follows that $g = g_3g_2'$ for some $g_3\theta g''\theta g_1'\theta g_1$ and $g_2'\theta g_2$. Thus, $Hg = Hg_3 \cdot g_2' \in \pi(Hg_1) \cdot \pi(Hg_2)$ as desired. A similar argument yields $\supseteq$.

This completes the proof of (1). It follows from (1), Lemmas 5.2.2 and 5.2.3 that

(2) $(G/H)^{**}$ is a cogroup and $G[H, \pi]$ is a polygroup.

In $(G/H)^{**}$,

$$Hg_1 \approx Hg_2 \iff Hg_1 \in \pi(Hg_2)H \iff Hg_1 = Hg_2'h$$

for some $g_2'\theta g_2$ and $h \in H$. Since θ is a conjugation and $H = \theta(e)$, $g_2'h\theta g_2'$, which implies that $g_1\theta g_2$. On the other hand $g_1\theta g_2$ implies that $(Hg_1)\pi(Hg)$, which yields $Hg_1 \approx Hg_2$. Thus, we obtain

(3) $Hg_1 \approx Hg_2$ if and only if $g_1\theta g_2$.

By (3), $\theta = \overline{\pi}$ (introduced in Lemma 5.2.5), so Lemma 5.2.5 yields $G//\theta \cong G[H, \pi]$ which completes the proof of theorem. ∎

As noted before the previous two results yield a factorization of double quotients.

Corollary 5.2.7. *$\mathcal{M} \in Q^2(Group)$ if and only if $\mathcal{M} \cong G[H, \pi]$ for some subgroup H of a group G and Utumi partition π on G/H.*

Denote the class of all polygroups derived from weak cogroups (respectively, cogroups) by the construction in Lemma 5.2.2 as $D(w\text{-cogroup})$ (respectively, $D(\text{cogroup})$). Then, Theorem 5.2.6 says

Corollary 5.2.8. *$Q^2(Group) \subseteq D(cogroup)$.*

We conclude by observing that all polygroups derived from weak cogroups are chromatic.

Theorem 5.2.9. *For every weak cogroup A, $\mathcal{M} = A/\approx$ is chromatic.*

Proof. Let $C = \{X \in M \mid X \neq \{e\}\}$ and for $X \in C$ let

$$C_X = \{(a,b) \in A^2 \mid a \in Xb\}.$$

It is not difficult to show that $\mathcal{V} = <A, C_x>_{x \in C}$ is a color scheme. As a sample of the argument consider

$$C_X \cap (C_Y \parallel X_Z) \neq \emptyset \text{ implies } C_X \subseteq C_Y \parallel C_Z.$$

Suppose that $a \in Xb$, $a \in Yc$ and $c \in Zb$ for some $c \in A$. Then, $a \in YZb$, from which it follows that $a \in ub$ for some $u \in y \cdot z$, where $y \in Y$ and $z \in Z$. Also, $a \in xb$ for some $x \in X$. Thus, condition (5) of Definition 5.2.1 yields $u \approx x$ and so $X \subseteq YZ$. Now, for any $(r,s) \in C_X$, $r \in Xs \subseteq YZs$, which gives $(r,s) \in C_Y \parallel C_Z$ as desired.

The other conditions are verified in a similar way. In particular, the argument used to show (4) also shows that for $X, Y, Z \neq \{e\}$, $X \in Y \cdot Z$ (in $\mathcal{M}_{\mathcal{V}}$) if and only if $C_X \subseteq C_Y \parallel C_Z$ if and only if $X \subseteq YZ$ (in $\mathcal{M}$). This is the key step in verifying the natural map from $\mathcal{M}_{\mathcal{V}}$ to $\mathcal{M}$ is an isomorphism. ∎

5.3 Conjugation lattice

Comer described a few elementary properties of the lattice of conjugation relations of a group. A decomposition of a group into double cosets as well as its decomposition into ordinary conjugation classes give examples of conjugation relations. Comer considered conjugations derivable from subsystems of polygroups and techniques for creating other conjugation from these. A lot of information about a group is coded into its conjugation lattice. In this section, we study conjugation relations on polygroups. The main reference for this section is [10].

For a polygroup P let $Conj(P)$ denote the collection of all conjugation relations on P and let $Conj_S(P)$ denote the collection of all special conjugations (see Definition 3.3.10). The *smallest conjugation relation* is the identity relation, denoted by δ_P and the *largest conjugation relation* is P^2 which is denotes by 1_P. Let 1_P^S denote the special conjugation relation which identifies all elements of P different from the identity e. When the polygroup P is understood δ, 1 and 1^S will be written instead of δ_P, 1_P and 1_P^S.

Definition 5.3.1. A partially ordered set $(L, \leq)$ is a *complete lattice* if every subset A of L has both a greatest lower bound (the infimum, also called the *meet*) and a least upper bound (the supremum, also called the *join*) in $(L, \leq)$. The meet of a and b is denoted by $a \wedge b$ and the join by $a \vee b$.

Definition 5.3.2. A non-empty subset I of a complete lattice $(L, \leq)$ is an *ideal*, if the following conditions hold:

(1) for every x, y in I, $x \vee y$ in I;
(2) for every x in I, $y \leq x$ implies that y is in I.

The smallest ideal that contains a given element a is a *principal ideal* and a is said to be a *principal element* of the ideal in this situation.

Proposition 5.3.3. *If P is a polygroup, then*

(1) *$Conj(P)$ forms a complete lattice whose join is the same as the join in the lattice of all equivalence relations on P;*
(2) *$Conj_S(P)$ is the principal ideal in $Conj(P)$ determined by 1^S.*

Proof. (1) It suffices to show, for every non-empty set S of $Conj(P)$ that the join $\bigvee S$ of S (in the lattice of all equivalence relations on P) is again a conjugation. Suppose that $z'(\bigvee S)z$ and $z \in x \cdot y$. Then, $z\theta_0 z_1 \theta_1 z_2 \ldots \theta_{n-1} z_n = z'$ for some $z_1, \ldots, z_{n-1} \in P$ and $\theta_0, \ldots, \theta_{n-1} \in S$. Because $\theta_0 \in Conj(P)$, $z_1 \in x_1 \cdot y_1$ for some $x_1 \theta_0 x$ and $y_1 \theta_0 y$ by Definition 3.3.10. Repeat for $\theta_1, \ldots, \theta_{n-1}$ to obtain $x_1, \ldots, x_n$ and $y_1, \ldots, y_n$ such that $x\theta_0 x_1 \theta_1 \ldots \theta_{n-1} x_n$, $y\theta_0 y_1 \theta_1 \ldots \theta_{n-1} y_n$ and $z_i \in x_i \cdot y_i$ for all $i \leq n$. Hence, $x_n(\bigvee S)x$, $y_n(\bigvee S)y$ and $z' \in x_n \cdot y_n$; so the second condition of Definition 3.3.10 holds for $\bigvee S$. The verification of the first condition is routine.

(2) It is obvious, since $\theta \in Conj(P)$ is special if and only if $\theta \leq 1^S$. ∎

There are situations when it is desirable to regard a conjugation on P as a partition of P instead of an equivalence relation. Partitions and equivalence relations will be interchanged freely. Partitions will be written in the form $\{A; B; \ldots\}$, where A, B, ... are the blocks of the partition. The largest special conjugation relation 1_P^S denotes the partition $\{\{e\}; P \backslash \{e\}\}$.

Notice that neither $Conj(P)$ nor $Conj_S(P)$ is a sublattice of the partition lattice, in general, because the intersection of two conjugation relations is not necessarily a conjugation relation.

In the following we give an example involving conjugations on $\mathbb{S}_3$.

Example 5.3.4. Let $\theta = \{\{0\}; \{(1\ 3)\}; \{(2\ 3), (1\ 2), (1\ 2\ 3), (1\ 3\ 2)\}\}$

and $\varphi = \{\{e\}; \{(1\ 2\}; \{(2\ 3), (1\ 3), (1\ 2\ 3), (1\ 3\ 2)\}\}$. Then, θ and φ are special conjugation relations on $\mathbb{S}_3$, but $\theta \cap \varphi = \{\{e\}; \{(1\ 3)\}; \{(1\ 2)\}; \{(2\ 3), (1\ 2\ 3), (1\ 3\ 2)\}\}$ is not a conjugation relation because $(2\ 3)(\theta \cap \varphi)(1\ 2\ 3)$ and $(1\ 2\ 3) = (1\ 2)((1\ 3)$ and $(2\ 3) \neq (1\ 2)((1\ 3)$.

The following lemma and the fact that $\mathbb{S}_3$ is generated by $\{(1\ 3), ((1\ 2)\}$ shows that for θ and φ above, $\theta \wedge \varphi$ is the identity conjugation relation.

Lemma 5.3.5. *If G is a group and $\theta \in Conj_S(G)$, then $H = \{x \in G \mid |\theta(x)| = 1\}$ is a subgroup of G.*

Proof. Clearly, $e \in H$ and H is closed under inverses. If $x, y \in H$, then

$$\theta(xy) \subseteq \theta(x)\theta(y) \subseteq \{xy\},$$

so H is closed under products also. ∎

Definition 5.3.6. An equivalence relation θ on a polygroup P is called a *congruence relation* if

    (1) θ is a regular relation (see Definition 2.5.1);
    (2) $x\theta y$ implies $x^{-1}\theta y^{-1}$, for all $x, y \in P$.

The lattice of all congruences on P is denoted by $Con(P)$.

In the following example, we give a conjugation relation associated with a subpolygroup H of a polygroup P.

Example 5.3.7.

    (1) For a subpolygroup H of P define a relation θ_H for $x, y \in P$ by

$$x\ \theta_H\ y \text{ if and only if } HxH = HyH.$$

    (2) If H is a subgroup of $Aut(P)$, define a relation θ^H for $x, y \in P$ by

$$x\ \theta^H\ y \text{ if and only if } \sigma(x) = y \text{ for some } \sigma \in H.$$

The relation θ_H is a conjugation relation and the relation θ^H is a special conjugation on P.

In Theorem 5.3.8, we show that the conjugations in Example 5.3.7 (1) include all congruence relations. It also shows that congruence relations correspond to normal subpolygroups.

Theorem 5.3.8. *Suppose that P is a polygroup.*

(1) *If $\theta \in Conj(P)$ and $N = \theta(e)$, then N is a subpolygroup of P and $\theta_N \subseteq \theta$.*

(2) *For an equivalence relation θ on P, $\theta \in Con(P)$ if and only if $\theta = \theta_N$ for some normal subpolygroup N of P.*

Proof. (1) It is clear that $e \in \theta(e)$ and $\theta(e)$ is closed under $^{-1}$. Now, suppose that $z \in x \cdot y$, where $x, y \in \theta(e)$. Then, $e\theta x$ and $x \in z \cdot y^{-1}$ so we conclude that $e \in z' \cdot y'$ for some $z'\theta z$ and $y'\theta y^{-1}$. But in a polygroup, $e \in z' \cdot y'$ gives $z' = (y')^{-1}$. So, $z\theta z' = (y')^{-1}\theta(y^{-1})^{-1} = y\theta e$ which shows that $z\theta e$. Thus, $\theta(e)$ is a subpolygroup of P. In order to show that $\theta_N \subseteq \theta$, suppose that $x\theta_N y$, i.e., $NxN = NyN$. Then, $x \in NxN = NyN \subseteq \theta(e)\theta(y)\theta(e) = \theta(y)$ since $\theta(e)$ is the identity of $P//\theta$. Thus, $x\theta y$ which completes the proof.

(2) It is straightforward to show that $\theta_N \in Con(P)$ whenever N is a normal subpolygroup of P.

Now, suppose that $\theta \in Con(P)$. In order to show $\theta \in Conj(P)$ it suffices to verify the second condition of Definition 3.3.10. Suppose that $z'\theta z$ and $z \in x \cdot y$. Then, $x \in z \cdot y^{-1}\overline{\theta} z' \cdot y^{-1}$ and $\theta \in Con(P)$. It follows that there exists $x' \in z' \cdot y^{-1}$ such that $x'\theta x$ which implies that $z' \in x' \cdot y$. Hence, $\theta \in Conj(P)$. Now, by part (1), $\theta_N \subseteq \theta$, where $N = \theta(e)$ is a subpolygroup of P. Suppose that $x\theta y$. Then, $e \in x \cdot x^{-1}$ and $x \cdot x^{-1}\overline{\theta} y \cdot x^{-1}$. So, there exists $z \in N$ with $z \in y \cdot x^{-1}$. Thus, $y \in z \cdot x \subseteq Nx$ which gives $x\theta_N y$. Hence, $\theta = \theta_N$. It remains that to show that N is normal. If $y \in Nx$, then $x\theta y$ which implies that $e \in x^{-1} \cdot x\overline{\theta}x^{-1} \cdot y$ by using the definition. Thus, for some $z\theta e$, $z \in x^{-1} \cdot y$ which gives $y \in x \cdot z \subseteq xN$. Therefore, $Nx \subseteq xN$. The other inclusion is similar. So, it follows that N is normal. ∎

By Example 5.3.7 (1), a conjugation relation θ_N is associated with every subpolygroup N of P, not just the normal ones. The following summarizes a few properties of this embedding.

Proposition 5.3.9. *Let P is a polygroup.*

(1) *The map $N \mapsto \theta_N$ embeds the lattice of subpolygroups of P into $Conj(P)$ as a join semilattice, note that a join semilattice is a partially ordered set which has a join for any non-empty finite subset.*

(2) *The image of the map in (1) has only $\theta_{\{e\}} = \delta$ in common with $Conj_S(P)$. In particular, $Con(P) \cap Conj_S(P) = \{\delta\}$.*

(3) *If G is a group, $Con(G)$ is a sublattice of $Conj(G)$.*

Proof. (1) Suppose that H and K are subpolygroups of P and $< H, K >$ is the subpolygroup generated by H and K. It suffices to show that $\theta_H \vee \theta_K = \theta_{<H,K>}$. If H and K are comparable, say $H \subseteq K$, then $< H, K >= K$ and $\theta_H \vee \theta_K = \theta_K = \theta_{<H,K>}$ clearly holds. Assume that H and K are not comparable. Then, $H \subseteq (\theta_H \vee \theta_K)(e)$ and $K \subseteq (\theta_H \vee \theta_K)(e)$. So, $< H, K >\subseteq (\theta_H \vee \theta_K)(e)$. By Theorem 5.3.8 (1), $\theta_{<H,K>} \leq \theta_H \vee \theta_K$. Since the other inclusion is clear, equality holds.

(2) If $\theta_H \in Conj_S(P)$, then $H = \{e\}$ by Theorem 5.3.8 and $\theta_H = \delta$.

(3) By a standard group theory argument $\theta_H \cap \theta_K = \theta_{H \cap K}$. Hence, $\theta_H \cap \theta_K$ is a conjugation relation which equals $\theta_H \wedge \theta_K$. ∎

By Theorem 5.3.8 (1) the block $\theta(e)$ of a conjugation relation θ is a subpolygroup. A new conjugation relation $\theta[\varphi]$ may be obtained from θ by replacing $\theta(e)$ block by the blocks of conjugation relation $\varphi \in Conj(\theta(e))$. More precisely,

Definition 5.3.10. Let P be a polygroup. For $\theta \in Conj(P)$ and $\varphi \in Conj(\theta(e))$, the φ-*split* of θ is an equivalence relation $\theta[\varphi]$ on P defined by

$$\theta[\varphi](x) = \begin{cases} \theta(x) \text{ if } \theta(x) \neq \theta(e) \\ \varphi(x) \text{ if } \theta(x) = \theta(e). \end{cases}$$

Proposition 5.3.11. $\theta[\varphi] \in Conj(P)$. *Moreover,* $P//(\theta[\varphi])$ *is isomorphic to* $(\theta(e)//\varphi)[P//\theta]$, *the polygroup extension of* $\theta(e)//\varphi$ *by* $[P//\theta]$ *(see Section 3.2).*

Proof. In order to verify the first statement it is enough to show that the product of two $\theta[\varphi]$-blocks is a union of $\theta[\varphi]$-blocks. Along the way we develop a rule for computing the product of two $\theta[\varphi]$-blocks from which the isomorphism is apparent. Let $\theta[\varphi] = \psi$ for short. The first two cases are obvious from the definition of $\theta[\varphi]$:

(1) $\psi(x)\psi(y) = \varphi(x)\varphi(y)$ if $\theta(x) = \theta(y) = \theta(e)$,
(2) $\psi(x)\psi(y) = \theta(x)\theta(y)$ if $\theta(x), \theta(y) \neq \theta(e)$.

When computing these products, replace $\theta(e)$ by $\{\varphi(x) \mid x \in \theta(e)\}$. For the other cases,

(3) $\psi(x)\psi(y) = \theta(y)$ if $\theta(x) = \theta(e) \neq \theta(y)$,
(4) $\psi(x)\psi(y) = \theta(x)$ if $\theta(y) = \theta(e) \neq \theta(x)$.

In order to verify (3), first note that $\varphi(x)\varphi(y) \subseteq \theta(e)\theta(y) = \theta(y)$. Now, suppose that $y'\theta y$. Then, $y' \in x \cdot z$ for some z. So, $y' \in x\theta(z) \subseteq \theta(e)\theta(z) = \theta(z)$. Then, $\theta(y) \cap \theta(z) \neq \emptyset$ so $\theta(z) = \theta(y)$ which gives $\theta(y) \subseteq x\theta(y) \subseteq$

$\theta(x)\theta(y)$ as desired. The proof of (4) is similar. Thus, $\theta[\varphi]$ is a conjugation relation. The isomorphism is established by comparing (1), (2), (3) and (4) with the definition of the product on the polygroup $(\theta(e)//\varphi)[P//\theta]$. ∎.

For $\varphi \leq \psi$ in $Conj(P)$ let $[\varphi, \psi]$ denote the universal $\{\theta \mid \varphi \leq \theta \leq \psi\}$ in $Conj(P)$. The map $\varphi \mapsto \theta[\varphi]$ immediately gives:

Corollary 5.3.12. *$Conj(\theta(e))$ is isomorphic to the interval $[\theta[\delta_{\theta(e)}], \theta]$ in $Conj(P)$.*

For a subset X of a polygroup such that $e \in X$ we let $X^* = X \setminus \{e\}$. There is one splitting of a conjugation relation θ that deserves special attention. Namely, for a conjugation relation θ let

$$\theta^S = \theta[1^S_{\theta(e)}],$$

where $1^S_{\theta(e)}$ is the unity element in $Conj_S(\theta(e))$. In other words, by Proposition 5.3.11, θ^S is a special conjugation relation obtained from θ by splitting $\theta(e)$ into the two classes: $\{e\}$ and $(\theta(e))^*$.

A few elementary properties of θ^S are given below.

Proposition 5.3.13. *Let P be a polygroup. For $\theta, \varphi \in Conj(P)$,*

(1) $\theta^S = \theta$ if θ is special and $\theta^S = \theta \cap 1^S$ if θ is not special;
(2) $\theta^S \leq \theta$;
(3) θ covers θ^S in $Conj(P)$ if θ is not special;
(4) $\varphi \leq \theta$ implies $\varphi^S \leq \theta^S$;
(5) θ is determined by θ^S and $\theta(e)$. Namely, $\theta = \theta_N | \theta^S$, a commuting join, where $N = \theta(e)$.

Proof. We prove (5). Suppose that $x\theta y$. If $\theta(x) \neq \theta(e)$, then $\theta^S(x) = \theta(x)$ so $x\theta_N x\theta^S y$ and if $\theta(x) = \theta(e)$ $(= N)$, then $x\theta_N y\theta^S y$. Therefore, $\theta \leq \theta_N | \theta^S$. ∎

Information about the structure of $Conj(P)$ can be obtained from Proposition 5.3.13 (5). For example, if G is an abelian group, every $\theta \in Conj(G)$ is a join of congruence relation and a special conjugation relation.

Lemma 5.3.14. *If h is a join retract of a lattice L onto an ideal L (i.e., $h : L \longrightarrow L$ satisfies $h(x \vee y) = h(x) \vee h(y)$, $h(x) \leq x$, $h(h(x)) = h(x)$ for all $x, y \in L$ and $h(L)$ is an ideal of L), then h is a homomorphism.*

Proof. Since h preserves order, $h(x \wedge y) \leq h(x) \wedge h(y)$. If $z \leq h(x) \wedge h(y)$, then $z \leq h(x) \leq x$ and $z \leq h(y) \leq y$ so $z \leq x \wedge y$. Hence, $z = h(z) \leq h(x \wedge y)$

because $z \leq h(x) \in h(L)$ implies $z \in h(L)$ and h fixes elements of $h(L)$. Thus, $h(x) \wedge h(y) = h(x \wedge y)$. ∎

Proposition 5.3.15. *The map* $\theta \mapsto \theta^S$ *is a lattice homomorphism of* $Conj(P)$ *onto* $Conj_S(P)$.

Proof. Applying Lemma 5.3.14, by Proposition 5.3.13, it suffices to show the map preserves joins. Since $\leq$ is preserved by Proposition 5.3.13 (4) we only need to show that $\theta \vee \varphi)^S \leq \theta^S \vee \varphi^S$ in $Conj(P)$. Suppose that $x(\theta \vee \varphi)^S y$ and $x, y \neq e$. Then, there exists a sequence $x = x_0, \ldots, x_n = y$ such that $x_0 \theta x_1 \varphi x_2 \ldots x_{i-1} \theta x_i \varphi x_{i+1} \ldots x_n$. If $x_j \theta x_{j+1}$ and $x_j, x_{j+1} \neq e$, then $x_j \theta^S x_{j+1}$ and similarly for φ. Hence, we may assume $x_i = e$ and $x_{i-1}, x_{i+1} \neq e$ for some i. Then, $x_{i-1} \in H = \theta(e)$ and $x_{i+1} \in K = \varphi(e)$. If $x_{i+1} \in H$, then $x_{i-1} \theta x_{i+1} \theta x_{i+2}$ and we may drop $e = x_i$ from the sequence. Hence we may assume $x_{i+1} \notin H$. Then, $e \notin x_{i-1} \cdot x_{i+1}$ because if so, $x_{i+1} = x_{i-1}^{-1} \in H$. Choose $x_i' \in x_{i-1} \cdot x_{i+1}$. Then, $x_i' \in H x_{i+1} \subseteq H x_{i+1} H \subseteq \theta(x_{i+1})$ and $x_i' \in x_{i-1} K \subseteq K x_{i-1} K \subseteq \varphi(x_{i-1})$. So, $x_{i-2} \varphi x_{i-1} \varphi x_i' \theta x_{i+1} \theta x_{i+2}$ which means we can shorten the sequence from x_0 to x_n and eliminate the term $x_i = e$. Repeating the above for all $x_i = e$ we obtain $x = x_0' \theta x_1' \varphi \ldots x_m' = y$ where all $x_i' \neq e$. Therefore, $(x, y) \in \theta^S \vee \varphi^S$. ∎

For $\theta, \varphi \in Conj(P)$ and $\varphi \subseteq \theta$, we define a conjugation relation $\theta // \varphi$ on $P // \varphi$ by

$$\varphi(x) \ (\theta // \varphi) \ \varphi(y) \ \Leftrightarrow \ x \theta y,$$

for all $x, y \in P$. The first part of the following theorem gives a lattice version of the first isomorphism theorem from group theory.

Theorem 5.3.16. *Let* $\varphi \in Conj(P)$.

(1) *The map* $\theta \mapsto \theta // \varphi$ *is an isomorphism of the interval* $[\varphi, 1]$ *in* $Conj(P)$ *onto* $Conj(P // \varphi)$.

(2) $\theta // \varphi$ *is special in* $Conj(P // \varphi)$ *if and only if* $\theta(e) = \varphi(e)$. *Moreover, the map in (1) is an isomorphism of the interval* $[\varphi, \overline{\varphi}]$ *in* $Conj(P)$ *onto* $Conj_S(P // \varphi)$.

(3) *The map* $\theta \mapsto \theta^S$ *is an isomorphism of* $[\varphi, \overline{\varphi}]$ *in* $Conj(P)$ *onto* $[\varphi^S, \overline{\varphi}^S]$ *in* $Conj_S(P)$.

(4) *If* φ *is not special, then the map* $\theta \mapsto \theta^S$ *is an isomorphism* $[\varphi, 1] \cong [\varphi^S, 1^S]$.

(5) *If* N *is a subpolygroup of* P, *then* $Conj_S(N) \cong [\theta_N[\delta], \theta_N^S]$, *a sublattice of* $Conj_S(P)$.

Proof. (1) It is a tedious but straightforward argument.

(2) We have

$$\theta//\varphi \text{ is special} \Leftrightarrow [\varphi(x) \, (\theta//\varphi) \, \varphi(e) \Rightarrow \varphi(x) = \varphi(e)]$$
$$\Leftrightarrow [x\theta e \Rightarrow x\varphi e]$$
$$\Leftrightarrow \theta(e) \subseteq \varphi(e).$$

But $\varphi(e) \subseteq \theta(e)$ always holds since $\varphi \subseteq \theta$. Since $\theta \in [\varphi, \overline{\varphi}]$ if and only if $\varphi(e) \subseteq \theta(e)$, the restriction of the map in (1) gives the desired isomorphism.

(3) If φ is special, $\theta^S = \theta$ for θ in $[\varphi, \overline{\varphi}] \subseteq Conj_S(P)$. Assume that $\varphi(e) \neq \{e\}$. Since $\theta \mapsto \theta^S$ is a lattice homomorphism it suffices to show the map is one to one and onto. Suppose that $\theta_1 \neq \theta_2$ in $[\varphi, \overline{\varphi}]$. Since $\theta_1(e) = \varphi(e) = \theta_2(e)$, there exists $x \notin \varphi(e)$ such that $\theta_1(x) \neq \theta_2(x)$. By Proposition 5.3.13 (1), $\theta_1^S(x) \neq \theta_2^S(x)$. So, the images are distinct and thus the map is one to one. Now, assume that θ is in $[\varphi^S, \overline{\varphi}^S]$. Define $\theta^+ = \{\varphi(e); \theta(x_1); \ldots\}$, where $\theta = \{e; (\varphi(e))^*; \theta(x_1); \ldots\}$. Since $(\theta^+)^S = \theta$ it suffices to show that θ^+ is a conjugation relation. If $x\theta^+y$, then it is clear that $x^{-1}\theta^+y^{-1}$ since $\varphi(e)^{-1} = \varphi(e)$ and $\theta(x_i)^{-1} = \theta(x_i^{-1})$ for all i. Thus, the first condition of Definition 3.3.10 holds. In order to verify the second condition, we need to show $\theta^+(x)\theta^+(y)$ is a union of θ^+-blocks. First, we show $\varphi(e)\theta(x_i) = \theta(x_i)$. This holds because $\varphi^S \subseteq \theta$ and $\theta(x_i) \neq \varphi(e)$ implies that $\theta(x_i)$ is a union of φ-blocks $\theta(x_i) = \varphi(x_i) \cup \ldots \cup \varphi(x_i') \cup \ldots$ and $\varphi(e)\varphi(x_i') = \varphi(x_i')$ for each component $\varphi(x_i')$. It remains to see that $\theta(x_i)\theta(x_j)$ is a union of θ^+-blocks. For this it suffices to show

$$(\varphi(e))^* \subseteq \theta(x_i)\theta(x_j) \Leftrightarrow e \in \theta(x_i)\theta(x_j).$$

Choose $x \in (\varphi(e))^*$. Then, $x \in x_i' \cdot x_j'$ for some $x_i'\theta x_i$, $x_j'\theta x_j$. Since $\theta(x) = \varphi^S(x) \subseteq \varphi^S(x_i')\varphi^S(x_j') = \varphi(x_i')\varphi(x_j')$, $e\varphi x$, and φ is a conjugation relation, $e \in (\varphi(x_i')\varphi(x_j') \subseteq \theta(x_i)\theta(x_j)$. The proof of the converse is similar. Thus, $\theta^+ \in Conj(P)$ which completes the proof of (3).

(4) It holds by an argument similar to (3).

(5) It follows from Corollary 5.3.12 and the observation that $\theta_N[\varphi]$ is special if and only if φ is special. ∎

In Theorem 5.3.16, if $\varphi \in Conj_S(P)$, then the homomorphism is, in general, not one to one.

Bibliography

[1] R. Ameri, *On categories of hypergroups and hypermodules*, J. Discrete Math. Sci. Cryptogr., 6 (2003) 121-132.

[2] H. Aghabozorgi, B. Davvaz and M. Jafarpour, *Solvable polygroups and derived subpolygroups*, Comm. Algebra, in press.

[3] I. Ben-Yaacov, *On the fine structure of the polygroup blow-up*, Arch. Math. Logic, 42 (2003) 649-663.

[4] P. Bonansinga, *Quasicanonical hypergroups. (Italian)*, Atti Soc. Peloritana Sci. Fis. Mat. Natur., 27 (1981) 9-17.

[5] P. Bonansinga, *Weakly quasicanonical hypergroups. (Italian)*, Atti Sem. Mat. Fis. Univ. Modena, 30 (1981) 286-298.

[6] P. Bonansinga and P. Corsini, *Sugli omomorfismi di semi-ipergruppi e di ipergruppi*, Boll. Un. Mat. Italy, 1-B (1982) 717-727.

[7] R.A. Borzoei, A. Hasankhani and H. Rezaei, *Some results on canonical, cyclic hypergroups and join spaces*, Ital. J. Pure Appl. Math., 11 (2002) 77-87.

[8] H. Campaigne, *Partition hypergroups*, Amer. J. Math., 6 (1940) 599-612.

[9] S.D. Comer, *Hyperstructures associated with character algebra and color schemes*, New Frontiers in Hyperstructures, Hadronic Press, (1996) 49-66.

[10] S.D. Comer, *Lattices of conjugacy relations*, Proceedings of the International Conference on Algebra, Part 3 (Novosibirsk, 1989), 31-48, Contemp. Math., 131, Part 3, Amer. Math. Soc., Providence, RI, 1992.

[11] S.D. Comer, *The representation of 3-dimensional cylindric algebras*, Algebraic logic (Budapest, 1988), 147-172, Colloq. Math. Soc. Janos Bolyai, 54, North-Holland, Amsterdam, 1991.

[12] S.D. Comer, *Multi-valued algebras and their graphical representations* (Preliminary draft), Math. and Comp. Sci. Dep. the Citadel. Charleston, South Carolina, 29409, July 1986.

[13] S.D. Comer, *The Cayley representation of polygroups, Hypergroups, other multivalued structures and their applications (Italian)* (Udine, 1985), 27-34, Univ. Studi Udine, Udine, 1985.

[14] S.D. Comer, *Polygroups derived from cogroups*, J. Algebra 89 (1984) 397-405.

[15] S.D. Comer, *Combinatorial aspects of relations*, Algebra Universalis, 18 (1984) 77-94.

[16] S.D. Comer, *A remark on chromatic polygroups*, Congr. Numer., 38 (1983) 85-95.

[17] S.D. Comer, *A new fundation for theory of relations*, Notre Dame J. Formal Logic, 24 (1983) 81-87.

[18] S.D. Comer, *Constructions of color schemes*, Acta Univ. Carolin. Math. Phys., 24 (1983) 39-48.

[19] S.D. Comer, *Combinatorial types*, Algebra, Combinatorics and Logic in Computer Science, Vol. I, II (Györ, 1983), 267–284, Colloq. Math. Soc. János Bolyai, 42, North-Holland, Amsterdam, 1986.

[20] S.D. Comer, *Extension of polygroups by polygroups and their representations using colour schemes*, Lecture notes in Meth., No 1004, Universal Algebra and Lattice Theory, (1982) 91-103.

[21] P. Corsini, *Rough sets, fuzzy sets and join spaces*, Honorary Volume dedicated to Prof. Emeritus J. Mittas, Aristotle's Univ. of Th., Fac. Engin., Math div., Thessaloniki, Greece, (2000) 65-72.

[22] P. Corsini, *Prolegomena of Hypergroup Theory*, Second edition, Aviani Editore, 1993.

[23] P. Corsini, *Join spaces, power sets, fuzzy sets*, Algebraic hyperstructures and applications (Iasi, 1993), 45–52, Hadronic Press, Palm Harbor, FL, 1994.

[24] P. Corsini, *(i.p.s.)* *Hypergroups of order 8*, Aviani Editore, (1989).

[25] P. Corsini, *Finite canonical hypergroups with partial scalar identities. (Italian)*, Rend. Circ. Mat. Palermo (2), 36 (1987) 205-219.

[26] P. Corsini, *(i.p.s.)* *ipergruppi di ordine 7*, Atti Sem. Mat. Fis. Un. Modena, 34, Modena, Italy, (1986) 199-216.

[27] P. Corsini, *Feebly canonical and 1-hypergroups*, Acta Univ. Carol., Math. Phys., 24 (1983) 49-56.

[28] P. Corsini, *Recent results in the theory of hypergroups (Italian)*, Boll. Un. Mat. Ital. A, (6) 2 (1983) 133-138.

[29] P. Corsini, *Contributo alla teoria degli ipergruppi*, Atti Soc. Pelor. Sc. Mat. Fis. Nat. Messina, Messina, Italy, (1980) 1-22.

[30] P. Corsini, *Sur les homomorphismes d'hypergroupes*, Rend. Sem. Univ., 52, Padova, Italy, (1974) 117-140.

[31] P. Corsini and V. Leoreanu, *Applications of Hyperstructures Theory*, Advanced in Mathematics, Kluwer Academic Publisher, (2003).

[32] P. Corsini and V. Leoreanu, *Hypergroups and binary relations*, Algebra Universalis, 43 (2000) 321-330.

[33] P. Corsini and V. Leoreanu, *About the heart of a hypergroup*, Acta Univ. Carolinae, 37 (1996) 17-28.

[34] I. Cristea and M. Ştefănescu, *Hypergroups and n-ary relations*, European J. Combin., 31 (2010) 780-789.

[35] B. Davvaz, *Polygroups and their properties*, in: Advances in Algebraic Structures: Proceedings of the International Conference on Algebra, Gadjah Mada University, Indonesia, 710 October 2010, World Scientific, (2010) 148-156.

[36] B. Davvaz, *Isomorphism theorems of polygroups*, Bull. Malays. Math. Sci. Soc. (2), 33 (2010) 385-392.

[37] B. Davvaz, *Applications of the* γ^**-relation to polygroups*, Comm. Algebra, 35 (2007) 2698-2706.

[38] B. Davvaz, *Groups in polygroups*, Iranian Journal of Mathematical Sciences and Informatics, 1(1) (2006) 25-31.

[39] B. Davvaz, *A new view of approximations in* H_v*-groups*, Soft Computing, 10 (2006) 1043 - 1046.

[40] B. Davvaz, *Rough subpolygroups in a factor polygroup*, Journal of Intelligent and Fuzzy Systems, 17 (2006) 613-621.

[41] B. Davvaz, *Elementary topics on weak polygroups*, Bull. Korean Math. Soc., 40 (2003) 1-8.

[42] B. Davvaz, *Fuzzy weak polygroups*, Algebraic hyperstructures and applications (Alexandroupoli-Orestiada, 2002), 127–135, Spanidis, Xanthi, 2003.

[43] B. Davvaz, *A brief survey of the theory of* H_v*-structures*, Algebraic hyperstructures and applications (Alexandroupoli-Orestiada, 2002), 39-70, Spanidis, Xanthi, 2003.

[44] B. Davvaz, *Rough polygroups*, Ital. J. Pure Appl. Math., 12 (2002) 91-96.

[45] B. Davvaz, *F-approximations in polygroups*, Int. Math. J., 2 (2002) 761-765.

[46] B. Davvaz, *On polygroups and weak polygroups*, Southeast Asian Bull. Math., 25 (2001) 87-95.

[47] B. Davvaz, *Polygroups with hyperoperators*, J. Fuzzy Math., 9 (2001) 815-823.

[48] B. Davvaz, *TL-subpolygroups of a polygroup*, Pure Math. Appl., 12 (2001) 137-145.

[49] B. Davvaz, *On polygroups and permutation polygroups*, Math. Balkanica (N.S.), 14 (2000) 41-58.

[50] B. Davvaz, *Weak polygroups*, Proceedings of the 28th Annual Iranian Mathematics Conference, Part 1 (Tabriz, 1997), 139–145, Tabriz Univ. Ser., 377, Tabriz Univ., Tabriz, 1997.

[51] B. Davvaz and F. Bardestani, *Hypergroups of type U on the right of size six*, Arab J. Sci. Eng., 36 (2011) 487-499.

[52] B. Davvaz and P. Corsini, *Generalized fuzzy polygroups*, Iranian J. Fuzzy Systems, 3 (2006) 59-75.

[53] B. Davvaz and M.A. Iranmanesh, *Fundamentals of Group Theory. (Persian)*, Yazd University, 2005.

[54] B. Davvaz and M. Karimian, *On the* γ_n*-complete hypergroups and* K_H *hypergroups*, Acta Math. Sin. (Engl. Ser.), 24 (2008) 1901-1908.

[55] B. Davvaz and M. Karimian, *On the* γ_n^* *complete hypergroups*, Euroupean J. Combinatorics, 28 (2007) 86-93.

[56] B. Davvaz and V. Leoreanu-Fotea, *Hyperring Theory and Applications*, International Academic Press, 115, Palm Harber, USA, 2007.

[57] B. Davvaz and V. Leoreanu-Fotea, *Applications of interval valued fuzzy n-ary polygroups with respect to t-norms (t-conorms)*, Comput. Math. Appl., 57 (2009) 1413-1424.

[58] B. Davvaz and V. Leoreanu-Fotea, *Binary relations on ternary semihypergroups*, Comm. Algebra, 38(10) (2010) 3621-3636.

[59] B. Davvaz and N.S. Poursalavati, *On polygroup hyperrings and representations of polygroups*, J. Korean Math. Soc., 36 (1999) 1021-1031.

[60] B. Davvaz and A.H. Sepahan-fard, *Small weak polygroups*, Transaction on applied mathematics and nonlinear models (TAMNOM), 1(1) (2008).

[61] B. Davvaz, A. Dehghan Nezad and A. Benvidi, *Chain reactions as experimental examples of ternary algebraic hyperstructures*, MATCH Communications in Mathematical and in Computer Chemistry, 65 (2011) 491-499.

[62] B. Davvaz, A. Dehghan Nezhad and A. Benvidi, *Chemical hyperalgebra: Dismutation reactions*, MATCH Communications in Mathematical and in Computer Chemistry, 67 (2012) 55-63.

[63] B. Davvaz, R. M. Santilli and T. Vougiouklis, *Studies of multivalued hyperstructures for the characterization of matter-antimatter systems and their extension*, Algebras Groups and Geometries, 28 (2011) 105-116.

[64] M. De Salvo, *Feebly canonical hypergroups*, Graphs, designs and combinatorial geometries (Catania, 1989), J. Combin. Inform. System Sci., 15 (1990) 133-150.

[65] M. De Salvo, K_H-*hypergroups. (Italian)*, Atti Sem. Mat. Fis. Univ. Modena, 31 (1982) 112-122.

[66] M. De Salvo, *Feebly canonical hypergroups*, J. Combin. Inform. System Sci., 15 (1990) 133-150.

[67] M. De Salvo and G. Lo Faro, *On the n^*-complete hypergroups*, Discrete Mathematics, 208/209 (1999) 177-188.

[68] A.P. Dietzman, *On the multigroups of complete conjugate sets of elements of a group*, C. R. (Doklady) Acad. Sci. URSS (N.S.) 49 (1946) 315-317.

[69] M. Dresher and O. Ore, *Theory of Multigroups*, Amer. J. Math., 60 (1938) 705-733.

[70] J.E. Eaton, *Theory of cogroups*, Duke Math. J., 6 (1940) 101-107.

[71] G. Falcone, *On finite strongly canonical hypergroups*, Pure Math. Appl., 11 (2000) 571-580.

[72] D. Freni, *Strongly transitive geometric spaces: applications to hypergroups and semigroups theory*, Comm. Algebra, 32 (2004) 969-988.

[73] D. Freni, *A new characterization of the derived hypergroup via strongly regular equivalences*, Comm. Algebra, 30 (2002) 3977-3989.

[74] D. Freni, *Une note sur le cur d'un hypergroupe et sur la clôture transitive β^* de β. (French) [A note on the core of a hypergroup and the transitive closure β^* of β]*, Riv. Mat. Pura Appl., 8 (1991) 153-156.

[75] M. Ghadiri, B. Davvaz and R. Nekouian, H_v-*Semigroup structure on F_2-offspring of a gene pool*, International Journal of Biomathematics, 5(4) (2012) 1250011 (13 pages).

[76] D.K. Harrison, *Double coset and orbit spaces*, Pacific J. Math., 80 (1979) 451-491.

[77] D.C. Higman, *Coherent configurations. I. Ordinary representation theory*, Geom. Dedicata, 4 (1975) 1-32.

[78] S. Hošková and J. Chvalina, *Discrete transformation hypergroups and transformation hypergroups with phase tolerance space*, Discrete Math., 308 (2008) 4133-4143.

[79] S.N. Hosseini, S.Sh. Mousavi and M.M. Zahedi, *Category of polygroup objects*, Bull. Iranian Math. Soc., 28 (2002) 67-86.

[80] T.W. Hungerford, *Algebra*, Graduate Texts in Mathematics, 73. Springer-Verlag, New York-Berlin, (1980).

[81] S. Ioulidis, *Polygroupes et certaines de leurs propriétés (French) [Polygroups and certain of their properties]*, Bull. Soc. Math. Grce (N.S.) 22 (1981) 95-104.

[82] A. Iranmanesh and A.H. Babareza, *Transposition hypergroups and complement hypergroups*, Algebraic hyperstructures and applications, 41-48, Taru Publ., New Delhi, 2004.

[83] A. Iranmanesh and M.N. Iradmusa, *The combinatorial and algebraic structure of the hypergroup associated to a hypergraph*, J. Mult.-Valued Logic Soft Comput., 11 (2005) 127-136.

[84] M. Jafarpour, H. Aghabozorgi and B. Davvaz, *On nilpotent and solvable polygroups*, Bull. Iranian Math. Soc., in press.

[85] J. Jantosciak, *Homomorphisms, equivalences and reductions in hypergroups*, Riv. Mat. Pura Appl., 9 (1991) 23-47.

[86] J. Jantosciak, *A brief survey of the theory of join spaces*, Algebraic hyperstructures and applications (Iasi, 1993), 1-12, Hadronic Press, Palm Harbor, FL, 1994.

[87] J. Jantosciak, *Transposition hypergroups: noncommutative join spaces*, J. Algebra 187 (1997) 97-119.

[88] J. Jantosciak and Ch.G. Massouros, *Strong identities and fortification in transposition hypergroups*, Algebraic hyperstructures and applications. J. Discrete Math. Sci. Cryptogr. 6 (2003) 169-193.

[89] M. Karimian and B. Davvaz, *On the γ-cyclic hypergroups*, Comm. Algebra, 34 (2006) 4579-4589.

[90] O. Kazanci, B. Davvaz and S. Yamak, *Fuzzy n-ary polygroups related to fuzzy points*, Comput. Math. Appl., 58 (2009) 1466-1474.

[91] G.J. Klir and T.A. Folger, *Fuzzy sets, uncertainty, and information*, Prentice-Hall International, Inc., 1988.

[92] M. Konstantinidou and K. Serafimidis, *Sur les filets des hypergroupes canoniques strictement réticulés. (French) [Threads of strictly lattice-ordered canonical hypergroups]*, Riv. Mat. Univ. Parma, (4) 13 (1987) 67-72.

[93] L. Konguetsof, T. Vougiouklis, M. Kessoglides and S. Spartalis, *On cyclic hypergroups with period*, Acta Univ. Carolin. Math. Phys., 28 (1987) 3-7.

[94] M. Koskas, *Groupoides, demi-hypergroupes et hypergroupes. (French)*, J. Math. Pures Appl., (9) 49 (1970) 155-192.

[95] M. Krasner, *A class of hyperrings and hyperfields*, Internat. J. Math. Math. Sci., 6 (1983) 307-311.

[96] V. Leoreanu, *The heart of some important classes of hypergroups*, Pure Math. Appl., 9 (1998) 351-360.

[97] V. Leoreanu, *New results on the hypergroups homomorphisms*, J. Inform. Optim. Sci., 20 (1999) 287-298.

[98] V. Leoreanu, *About the simplifiable cyclic semihypergroup*, Italian J. Pure Appl. Math., 7 (2000) 69-76.

[99] V. Leoreanu-Fotea and B. Davvaz, *n-hypergroups and binary relations*, European Journal of Combinatorics, 29(5) (2008) 1207-1218.

[100] R.C. Lyndon, *Relation algebras and projective geometries*, Michigan Math. J., 8 (1961) 21-28.

[101] R.D. Maddux, *Finite polygroups*, Math 492, Spring (2002).

[102] R. Maddux, *Embedding modular lattices into relation algebras*, Algebra Universalis, 12 (1981) 242-246.

[103] F. Marty, *Sur une généralization de la notion de groupe*, 8th Congress Math. Scandenaves, Stockholm, (1934) 45-49.

[104] Ch.G. Massouros, *Canonical and join hypergroups*, An. Ştiinţ. Univ. Al. I. Cuza Iaşi. Mat. (N.S.), 42 (1996) 175-186.

[105] Ch.G. Massouros, *Quasicanonical hypergroups*, Algebraic hyperstructures and applications, Proc. 4th Int. Congr., Xanthi- Greece 1990, 129-136.

[106] J.R. McMullen and J.F. Price, *Reversible hypergroups*, Rend. Sem. Mat. Fis. Milano, 47 (1977) 67-85.

[107] J.R. McMullen and J.F. Price, *Duality for finite abelian hypergroups over splitting fields*, Bull. Austral. Math. Soc., 20 (1979) 57-70.

[108] R. Migliorato, *Canonical v-hypergroups and ω-hypergroups*, J. Discrete Math. Sci. Cryptography, 6 (2003) 245-256.

[109] R. Migliorato, *On the complete hypergroups*, Riv. Di Mat. Pura e Appl., 12 (1994) 21-31.

[110] R. Migliorato, *Semi-ipergruppi e ipergruppi n-completi*, Ann. Sci. Univ. Clermont II, Ser Math, 23 (1986) 99-123.

[111] J. Mittas, *Hypergroupes canoniques valus et hypervaluós–hypergroupes fortement et supérieurement canoniques. (French) [Valued and hypervalued canonical hypergroups—strongly and predominantly canonical hypergroups]*, Bull. Soc. Math. Grce (N.S.) 23 (1982) 55-88.

[112] J. Mittas, *Hypergroupes canoniques*, Math. Balkanica, Beograd 2 (1972) 165-179.

[113] J. Mittas, *Contributions à la théorie des hypergroupes, hyperanneaux et hypercorps hypervalués. (French)* C. R. Acad. Sci. Paris Sr. A-B 272 (1971) A3-A6.

[114] J. Mittas, *Hypergroupes canoniques valués et hypervalués. (French)*, Math. Balkanica, 1 (1971) 181-185.

[115] J. Mittas, *Hypergroupes canoniques hypervalués. (French)*, C. R. Acad. Sci. Paris Sér. A-B 271 (1970) A4-A7.

[116] J. Mittas, *Sur une classe d'hypergroupes commutatifs. (French)*, C. R. Acad. Sci. Paris Sér. A-B 269 (1969) A485-A488.

[117] C. Pelea, *About a category of canonical hypergroups*, Ital. J. Pure Appl. Math., 7 (2000) 157-166.

[118] W. Prenowitz, *Spherical geometries and multigroups*, Canadian J. Math., 2 (1950) 100-119.

[119] W. Prenowitz, *Projective geometries as multigroups*, Amer. J. Math., 65 (1943) 235-256.

[120] W. Prenowitz, *A contemporary approach to classical geometry*, Amer. Math. Monthly, 68(1) (1961) part II.

[121] W. Prenowitz, *Partially ordered fields and geometries*, Amer. Math. Monthly, 53 (1946) 439-449.

[122] W. Prenowitz, *Descriptive geometries as multigroups*, Trans. Amer. Math. Soc., 59, (1946) 333-380.

[123] W. Prenowitz and J. Jantosciak, *Geometries and join spaces*, J. Reine Angew. Math., 257 (1972) 100-128.

[124] W. Prenowitz and J. Jantosciak, *Join Geometries*, Springer-Verlag, UTM, (1979).

[125] J.S. Rose, *A Course on Group Theory*, Cambridge University Press, (1978).

[126] R. Rota, *Sugli iperanelli moltiplicativi*, Rend. Di Mat., Series VII (4) 2, (1982) 711-724.

[127] I.G. Rosenberg, *Hypergroups induced by paths of a direct graph*, Ital. J. Pure Appl. Math., 4 (1998) 133-142.

[128] I.G. Rosenberg, *Hypergroups and join spaces determined by relations*, Ital. J. Pure Appl. Math., 4 (1998) 93-101.

[129] R.L. Roth, *Character and conjugacy class hypergroups of a finite group*, Ann. Mat. Pura Appl., (4) 105 (1975) 295-311.

[130] R.L. Roth, *On derived canonical hypergroups*, Riv. Mat. Pura Appl., 3 (1988) 81-85.

[131] J.J. Rotman, *An Introduction to the Theory of Groups*, Fourth edition, Graduate Texts in Mathematics, 148. Springer-Verlag, New York, (1995).

[132] D. Schweigert, *Congruence relations of multialgebras*, Discrete Math., 53 (1985) 249-253.

[133] K. Serafimidis, *Sur les hypergroupes canoniques ordonnés et strictement ordonnés. (French) [Ordered and strictly ordered canonical hypergroups]*, Rend. Mat., (7) 6 (1986) 231-238.

[134] K. Serafimidis, M. Konstantinidou and J. Mittas, *Sur les hypergroupes canoniques strictement réticulés. (French) [On strictly lattice-ordered canonical hypergroups]*, Riv. Mat. Pura Appl., 2 (1987) 21-35.

[135] M. Ştefănescu, *Some interpretations of hypergroups*, Bull. Math. Soc. Sci. Math. Roumanie (N.S.), 49(97) (2006) 99-104.

[136] S. Spartalis, *On reversible H_v-groups*, Algebraic hyperstructures and applications (Iasi, 1993), 163-170, Hadronic Press, Palm Harbor, FL, (1994).

[137] Y. Sureau, *On structure of cogroups*, Discrete Mathematics, 155 (1996) 243-246.

[138] Y. Sureau, *Contribution a la theorie des hypergroupes operant transitivement sur un ensemble*, These de Doctorate d'Etat, Universite de Clermont II, (1980).

[139] M. Suzuki, *Group theory I*, Translated from the Japanese by the author, Grundlehren der Mathematischen Wissenschaften [Fundamental Principles of Mathematical Sciences], 247, Springer-Verlag, Berlin-New York, (1982).

[140] G. Tallini, *On Steiner hypergroups and linear codes*, Hypergroups, other multivalued structures and their applications (Italian) (Udine, 1985), 87-91, Univ. Studi Udine, Udine, (1985).

[141] Y. Utumi, *On hypergroups of groups right cosets*, Osaka Math. J., 1 (1949) 73-80.

[142] J.C. Varlet, *Remarks on distributive lattices*, Bull. de l'Acad. Polonnaise des Sciences, Serie des Sciences Math., Astr. et Phys., XXIII, n. 11 (1975)

1143-1147.

[143] T. Vougiouklis, *Convolutions on WASS hyperstructures*, Combinatorics (Rome and Montesilvano, 1994). Discrete Math., 174 (1997) 347-355.

[144] T. Vougiouklis, H_v-*groups defined on the same set*, Discrete Math., 155 (1996) 259-265.

[145] T. Vougiouklis, *Some results on hyperstructures*, Contemporary Math., 184 (1995) 427-431.

[146] T. Vougiouklis, *A new class of hyperstructures*, J. Combin. Inform. System Sci., 20 (1995) 229-235.

[147] T. Vougiouklis, *Hyperstructures and their Representations*, Hadronic Press, Inc, 115, Palm Harber, USA, 1994.

[148] T. Vougiouklis, *Representations of hypergroups by generalized permutations*, Algebra Universalis, 29 (1992) 172-183.

[149] T. Vougiouklis, *The fundamental relation in hyperrings. The general hyperfield*, Algebraic hyperstructures and applications (Xanthi, 1990), 203-211, World Sci. Publishing, Teaneck, NJ, (1991).

[150] T. Vougiouklis, *The very thin hypergroups and the S-construction*, Combinatorics 88, Vol. 2 (Ravello, 1988), 471-477, Res. Lecture Notes Math., Mediterranean, Rende, (1991).

[151] T. Vougiouklis, *Groups in hypergroups*, Annals Discrete Math., 37 (1988) 459-468.

[152] T. Vougiouklis, *Representations of hypergroups by hypermatrices*, Riv. Mat. Pura Appl., 2 (1987) 7-19.

[153] T. Vougiouklis and S. Spartalis, *P-cyclic hypergroups with three characteristic elements*, Combinatorics 86 (Trento, 1986), 421-426, Ann. Discrete Math., 37, North-Holland, Amsterdam, (1988).

[154] T. Vougiouklis, *Representation of hypergroups. Hypergroup algebra*, Convegno: Ipergruppi, str. mult. appl. Udine (1985) 59-73.

[155] T. Vougiouklis, *Cyclicity in a special class of hypergroups*, Acta Univ. Carolin.–Math. Phys., 22 (1981) 3-6.

[156] H.S. Wall, *Hypergroups*, Amer. J. Math., 59 (1937) 77-98.

[157] C.N. Yatras, *Subhypergroups of M-polysymmetrical hypergroups*, Algebraic hyperstructures and applications (Iaşi, 1993), 123-132, Hadronic Press, Palm Harbor, FL, (1994).

[158] C.N. Yatras, *M-polysymmetrical hypergroups*, Riv. Mat. Pura Appl., 11 (1992) 81-92.

[159] C.N. Yatras, *Homomorphisms in the theory of the M-polysymmetrical hypergroups and monogene M-polysymmetrical hypergroups*, Proceedings of the Workshop on Global Analysis, Differential Geometry and Lie Algebras (Thessaloniki, 1995), 155-165, BSG Proc., 1, Geom. Balkan Press, Bucharest, (1997).

[160] M.M. Zahedi, M. Bolurian and A. Hasankhani, *On polygroups and fuzzy subpolygroups*, J. Fuzzy Math. 3 (1995) 1-15.

[161] M.M. Zahedi, L. Torkzadeh and R.A. Borzooei, *Hyper I-algebras and polygroups*, Quasigroups Related Systems, 11 (2004) 103-113.

Index